Quantitative Evaluation of Measurement Uncertainty

Quantitative Evaluation of Measurement Uncertainty: Theory and Practice provides comprehensive coverage of dynamic, direct, indirect, and combined measurements, giving a complete perspective on measurement uncertainty. It features pioneering approaches to evaluating uncertainty in dynamic measurements.

Responding to the critical need for standardized approaches to uncertainty evaluation, this book provides a clear pathway to ensuring that measurement results include properly calculated uncertainty intervals with genuine coverage probabilities. With detailed practical examples spanning physics, chemistry, electrical engineering, and mechanical engineering, readers will gain a thorough understanding of both statistical and non-statistical methodologies for uncertainty analysis.

Bridging theoretical principles with practical applications, this book will provide engineers, researchers, and graduate students with the essential tools to assess and communicate measurement uncertainty with confidence.

Quantitative Evaluation of Measurement Uncertainty

Theory and Practice

Oleksandr Vasilevskyi
Michael Cullinan

CRC Press
Taylor & Francis Group
Boca Raton London New York

CRC Press is an imprint of the
Taylor & Francis Group, an **informa** business

First edition published 2026
by CRC Press
2385 NW Executive Center Drive, Suite 320, Boca Raton FL 33431

and by CRC Press
4 Park Square, Milton Park, Abingdon, Oxon, OX14 4RN

CRC Press is an imprint of Taylor & Francis Group, LLC

ISBN: 978-1-041-24599-5 (hbk)
ISBN: 978-1-041-24600-8 (pbk)
ISBN: 978-1-003-74270-8 (ebk)

DOI: 10.1201/9781003742708

Typeset in Times
by SPi Technologies India Pvt Ltd (Straive)

Contents

About the Authors

Oleksandr Vasilevskyi (Member, IEEE) is currently a senior researcher at the Walker Department of Mechanical Engineering, the Cockrell School of Engineering, The University of Texas at Austin, USA. He received his PhD degree in information and measuring systems from Vinnytsia National Technical University, Ukraine, in 2006, and his Doctor of Engineering Sciences (Higher Doctorate) degree in standardization, certification, and metrological equipment from Lviv Polytechnic National University, Ukraine, in 2015. Prof. Vasilevskyi was conferred the academic title of Professor (Full) in metrology and industrial automation by the Ministry of Education and Science of Ukraine based on the decision of the Academic Council of Vinnytsia National Technical University in 2017. Prof. Vasilevskyi is the recipient of many awards for his research and teaching, including the Laureate of the Prize of the Verkhovna Rada of Ukraine for Young Scientists in 2019, the Badge of the Ministry of Education and Science of Ukraine for "Scientific and Educational Achievements" in 2020, and the Badge of the Ministry of Education and Science of Ukraine for "Excellence in Education of Ukraine" in 2019.

Michael Cullinan (Member, IEEE) is currently an associate professor and the Associate Department Chair for graduate studies at the Walker Department of Mechanical Engineering, The University of Texas at Austin, USA. He received his PhD degree in mechanical engineering from Massachusetts Institute of Technology (MIT), USA, in 2011. He is also the Director of the UT Austin's Semiconductor Science and Engineering Program. Dr. Cullinan is the recipient of many awards for his research and teaching, including American Society for Precision Engineering Early Career Award in 2021; the ASME Kornel F. Ehmann Manufacturing Medal in 2020; the Outstanding Young Manufacturing Engineer Award from the Society of Manufacturing Engineers in 2016; the Rising Star Award from the Sensors Expo and Conference in 2017; the multiple Best Poster Awards from American Society for Precision Engineering in 2017 and 2018; and the Outstanding Teaching by an assistant professor Award in 2017 and an associate or full professor from the Department of Mechanical Engineering, The University of Texas at Austin, in 2024.

Introduction

In the field of measurements and testing, it is essential not only to obtain a result but also to evaluate reliability. For this purpose, the uncertainty associated with the result must be quantitatively assessed. Without such an evaluation, it is impossible to compare results with each other or with reference values provided in specifications and standards. This textbook is based on the idea that the quantitative evaluation of measurement uncertainty is a fundamental component of modern measurement practice. It introduces a clear, practical, and widely accepted framework for uncertainty analysis, which is necessary for ensuring the credibility and comparability of measurement results.

In modern measurement science, the concept of measurement uncertainty has become a fundamental element for interpreting and evaluating results. While error and error analysis have traditionally been used in measurements, it is now generally accepted that even after accounting for known errors and applying corrections, some degree of uncertainty remains. This residual uncertainty reflects the inevitable doubt about how closely the result represents the true value of the quantity being measured. Understanding and quantifying this uncertainty is essential for ensuring the reliability and comparability of measurement results.

Just as the widespread adoption of the International System of Units (SI) has brought consistency to scientific and technological measurements, a globally harmonized approach to the evaluation and expression of measurement uncertainty is essential for the accurate interpretation and effective use of measurement results across various fields—including science, industry, medicine, and international trade. In today's globalized world, methods for evaluating and presenting uncertainty must be standardized worldwide, ensuring that measurements performed in different countries can be reliably compared and trusted.

The quantity used to represent uncertainty must be internally consistent: it should be directly derived from the components that form it and should not depend on how these components are grouped or broken down into subcomponents. Additionally, it should be possible to use the uncertainty of one result as a component in evaluating the uncertainty of another result in which the first result is used. Furthermore, in many industries, commerce, and areas related to health and safety, it is often necessary to present measurement results with an interval that most likely contains a significant portion of the distribution of values that can reasonably characterize the measured quantity. Therefore, the ideal methodology for evaluating and presenting uncertainty should provide such an interval, including a coverage probability or confidence level that genuinely corresponds to it.

The primary objective of this textbook is to provide higher education students and professionals with the skills and knowledge required to assess measurement result uncertainty and present research findings based on the concept of measurement uncertainty.

1 Concept of Evaluating Measurement Result Uncertainty

1.1 DEFINITIONS OF MEASUREMENT UNCERTAINTY

The term uncertainty refers to doubt, and thus, in a broader sense, "measurement uncertainty" means doubt regarding the reliability of the measurement result. The formal definition of the term "measurement uncertainty" is as follows: measurement uncertainty is a parameter associated with the measurement result that characterizes the dispersion of values that can be reasonably attributed to the measured quantity.[1,2]

Uncertainty of a measurement result reflects the lack of exact knowledge of the value of the measured quantity. After applying corrections for known systematic effects, the measurement result remains only an estimate of the value of the measured quantity due to uncertainties arising from random effects and the imperfect correction of systematic effects. The measurement result (after applying the correction) can be very close to the true value of the measured quantity (and therefore has a very small error), even if it has a large uncertainty. Thus, the uncertainty of the measurement result should not be confused with the unknown error that remains.

Since the exact values of the error components of a measurement result are unknown and unknowable, the uncertainties associated with the random and systematic effects causing the error can only be estimated. However, even if the estimated uncertainties are small, there is no guarantee that the measurement error will also be small, because when determining corrections or assessing the incompleteness of knowledge, some systematic effect may not be considered if it has not been recognized. Thus, the uncertainty of a measurement result does not necessarily indicate the likelihood that the result is close to the true value of the measured quantity; it is simply an estimate of the closeness of the measurement result to the best value corresponding to the knowledge available at that time.

Measurement uncertainty expresses the fact that, for a given measurand and a given measurement result, there is no single true value, but rather an infinite number of values scattered around the result. These values are consistent with all observations and data, as well as with the knowledge of the physical world, and may be attributed to the measurand with varying degrees of confidence. Figure 1.1 illustrates the concept of measurement uncertainty. From this figure, it becomes clear why the primary focus is placed on uncertainty rather than on error.[3]

The exact value of the measurement error is generally unknown. All that can be done is to estimate the values of the input quantities, considering corrections for known systematic effects along with their standard uncertainties (evaluated as

DOI: 10.1201/9781003742708-1

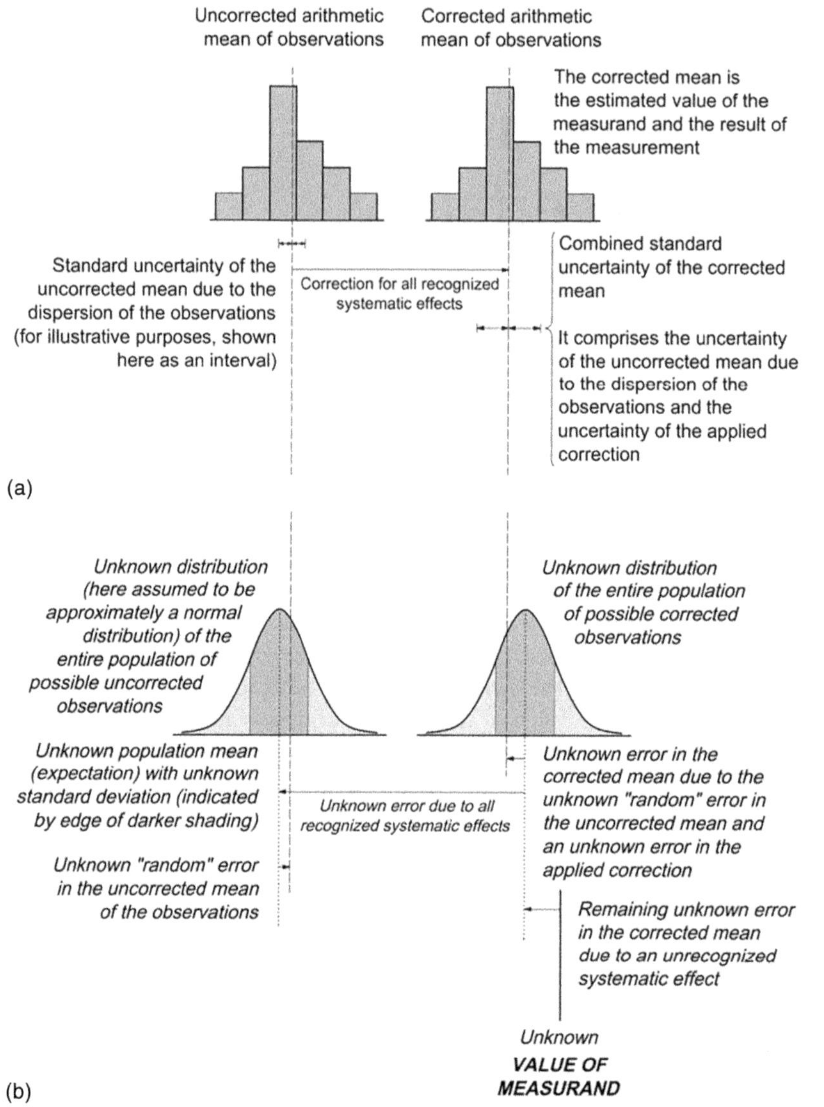

FIGURE 1.1 Graphical illustration of the meaning of measurement uncertainty. (a) Concepts based on observable quantities. (b) Ideal concepts based on unknowable quantities.

standard deviations), which may arise from both unknown probability distributions—sampled through repeated observations—and from subjective or a priori distributions based on all available information. The measurement result is then calculated using the estimated values of the input quantities and the combined uncertainty of that result, relying on the standard uncertainties, provided there is strong confidence that all these steps have been correctly carried out and all significant systematic effects have been accounted for. In that case, the measurement result may be considered a reliable estimate of the value of the measurand, and its combined uncertainty a reliable measure of its possible error.

In practice, there are many possible sources of measurement uncertainty, including:

1. incomplete definition of the quantity;
2. imperfect realization of the definition of the quantity;
3. non-representative sampling —the obtained value may not reflect the quantity;
4. inaccurate knowledge of the environmental influences on the measurement or imperfect measurement of environmental conditions;
5. subjective systematic error of the operator when reading analog instruments;
6. finite resolution of the instrument or threshold of sensitivity;
7. inaccurate values assigned to reference standards, certified reference materials, or substances;
8. inaccurate values of constants and other parameters obtained from external sources and used in the data processing algorithm;
9. approximations and assumptions used in the measurement method and measurement procedure;
10. variations in repeated observations of the quantity under apparently identical conditions.

These sources are not necessarily independent, and some of the sources from (1) to (9) may contribute to source (10). Naturally, an unknown systematic effect cannot be included in evaluating the measurement uncertainty, but it contributes to the measurement error.

The introduction of the concept of "measurement uncertainty" is necessary for a uniform and simplified assessment of measurement reliability since its definition is based on the obtained measurement results, known measurement conditions, and characteristics of the equipment used rather than on the unknown true value of the measurand.

1.2 SYSTEMATIZATION OF MEASUREMENT UNCERTAINTIES

Measurement uncertainties can be systematized according to the following criteria: *by methods of evaluation* and *by forms of expression* (Figure 1.2).

According to the evaluation methods, all measurement uncertainties are classified as Type A and Type B. The purpose of distinguishing between Type A and Type B is to illustrate two different approaches to evaluating components of uncertainty, and this classification is used solely for convenience in discussion; it is not intended to imply that there is a difference in the nature of the components resulting from the two types of evaluation. Both types of evaluation are based on probability distributions, and the uncertainty components of each type are quantitatively characterized by variance or standard deviation.

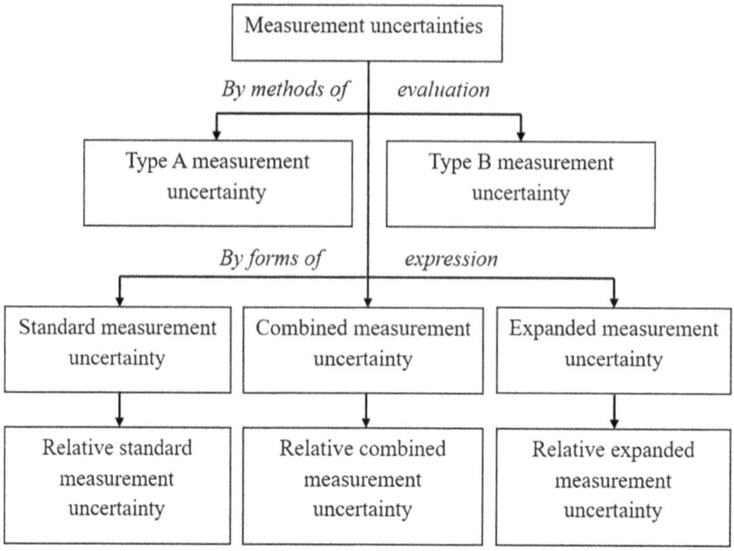

FIGURE 1.2 Systematization of measurement uncertainties.

Type A evaluation is applied to uncertainties that are subject to repeated measurements and to which statistical methods can be applied.

A posteriori evaluation is carried out based on the results of measurement, inspection, or testing, and is only possible when repeated observations of the measured quantity are performed. Such measurements can be conducted in two modes:

1. under repeatability conditions (for evaluating and minimizing the measurement uncertainty caused by random effects);
2. by intentionally varying one of the observation conditions to obtain observable variability in the results (for evaluating and minimizing the uncertainty of measurement results caused by the variable component of an uneliminated known systematic effect).

As a result of processing repeated observations using mathematical statistics, it is possible to obtain a measure of their dispersion around the estimate of the expected value (reference value), which is taken as the measurement result. The measure of dispersion of the observation results is expressed by the experimental standard deviation, which is referred to as **the Type A standard uncertainty**.

Type B evaluation is applied to uncertainties for which statistical methods cannot be used. In such cases, other known approaches are utilized.

That is, a priori evaluation of the components of measurement uncertainty must be performed when repeated observations are not carried out or cannot be conducted. In this case, one should rely on information obtained from previous measurements, the physical properties of the quantity, specification data of the measuring instrument, or reference materials.

As for the systematization of uncertainties **by forms of expression**, they are classified as standard, combined, expanded, relative standard, relative combined, and relative expanded uncertainties (see Figure 1.2).

The standard measurement uncertainty is the measurement uncertainty expressed as a standard deviation (root-mean-square deviation).

The combined measurement uncertainty is the measurement uncertainty obtained by combining all individual standard measurement uncertainties associated with the measured quantity.

The expanded measurement uncertainty defines an interval about the measurement result that is expected to encompass a large fraction of the distribution of values that could reasonably be attributed to the measurand.

Relative standard measurement uncertainty is the ratio of the standard measurement uncertainty to the estimate of the **input** quantity.

Relative combined measurement uncertainty is the ratio of the combined measurement uncertainty to the estimate of the **output** quantity.

Relative expanded measurement uncertainty is the ratio of the expanded measurement uncertainty to the estimate of the **output** quantity.

1.3 METHODS FOR EVALUATING STANDARD MEASUREMENT UNCERTAINTIES

The measurement uncertainty evaluation that characterizes the measurement method's accuracy is called a priori evaluation. It is determined in the following cases:

1. during the development of the measurement procedure to regulate the prescribed uncertainty under all measurement conditions specified by the procedure;
2. in the absence of a procedure or a prescribed uncertainty value, it is evaluated before the measurement to estimate the possible uncertainty.

Based on all available information about the causes and sources of uncertainty, individual components of Type B measurement uncertainty, combined measurement uncertainty, and expanded measurement uncertainty are calculated. The foundation of **a priori** measurement uncertainty evaluation is the theory of probability, which makes it possible to study and describe the distribution laws of random variables.

The measurement uncertainty evaluation for specific measurement results is **a posteriori** (the Type A standard measurement uncertainty); it is performed directly after the measurement, under specific conditions, using a defined procedure and specific types of measuring instruments. The foundation of a posteriori uncertainty evaluation is the methods of mathematical statistics, which can be used to assess the degree of dispersion in the results of repeated observations.

1.3.1 Evaluation of Type A Measurement Uncertainty

The experimental variance, which characterizes the measurement uncertainty component obtained through Type A evaluation, is derived from a series of repeated observations and represents a statistical variance estimate. The experimental standard deviation is calculated as the positive square root of the variance, denoted by u_A, and for convenience, it is referred to as the Type A standard measurement uncertainty.

Assessment of Type A standard measurement uncertainty components is based on frequency distributions. Therefore, to evaluate the Type A standard measurement uncertainty, it is necessary to perform n independent observations of the measurand q under repeatability conditions.

In most cases, the best available estimate of the expected value (mathematical expectation) μ_s of a randomly varying quantity q is the arithmetic mean or average value $\bar{q}$ calculated from n repeated observations (1.1):

$$\bar{q} = \frac{1}{n}\sum\nolimits_{k=1}^{n} q_k. \tag{1.1}$$

The experimental standard deviation, which characterizes the variability of the values q_k, or more precisely, their variance σ^2 relative to the mean value $\bar{q}$, is calculated using the formula (1.2):

$$u_A(q_k) = \sqrt{\frac{\sum_{k=1}^{n}(q_k - \bar{q})^2}{n-1}}. \tag{1.2}$$

Since the mean value $\bar{q}$ is accepted as the result of multiple measurements, it is important to estimate its variance $\sigma^2(\bar{q})$. The best estimate $\sigma^2(\bar{q}) = \sigma^2/n$ of the variance of the mean value $u_A^2(\bar{q})$ is expressed as (1.3):

$$u_A^2(\bar{q}) = \frac{u_A^2(q_k)}{n}. \tag{1.3}$$

The experimental variance of the mean $u_A^2(\bar{q})$ and the experimental standard deviation of the mean $u_A(\bar{q})$, which is the positive square root of the variance estimate $u_A^2(\bar{q})$, quantitatively characterize how well $\bar{q}$ estimates the expected value μ_s of the quantity q.

Considering Equations (1.2) and (1.3), the experimental standard deviation of the mean $u_A(\bar{q})$ is calculated using the following formula (1.4):

$$u_A(\bar{q}) = \sqrt{\frac{\sum_{k=1}^{n}(q_k - \bar{q})^2}{n(n-1)}}. \tag{1.4}$$

To reduce the standard uncertainty of the measurement result, it is advisable to perform measurements with multiple observations. The number of observations n should be increased until the mean value $\bar{q}$ provides a reliable estimate of the expectation μ_s of the random variable q, and the standard deviation of the mean $u_A^2(\bar{q})$ ensures a reliable estimate of the variance $\sigma^2(\bar{q})$.

The standard deviation of the estimate of the standard deviation of the arithmetic mean $\sigma[u_A(\bar{q})]$, for a normal distribution of q, is determined by the following equation (1.5):

$$\sigma[u_A(\bar{q})] = \sigma(\bar{q})\sqrt{1 - \frac{2\Gamma^2\left(\frac{n}{2}\right)}{(n-1)\Gamma^2\left(\frac{n-1}{2}\right)}}, \tag{1.5}$$

where $\Gamma(z)$ is the gamma function, which is defined by the expression (1.6):

$$\Gamma(z) = \int_0^\infty t^{z-1}e^{-t}dt, \text{for } z > 0. \tag{1.6}$$

The dependence of the ratio of the standard deviation of the experimental standard deviation of the arithmetic mean $u_A(\bar{q})$ to the standard deviation $\sigma(\bar{q})$ on the number of observations, obtained from expression (1.5), is approximately described by the following expression (1.7):

$$\frac{\sigma[u_A(\bar{q})]}{\sigma(\bar{q})} \approx \frac{1}{\sqrt{2(n-1)}}. \tag{1.7}$$

If the existence of such a deviation is acknowledged, and its potential magnitude is expected to be significant, it may be represented using a simple probability distribution. This distribution should reflect the available information that supports the assumption that the deviation exists and could be substantial. When probability is interpreted as a degree of belief in the occurrence of an event, the contribution of this systematic effect can be included in the combined standard measurement uncertainty by evaluating it as a standard uncertainty based on an a priori probability distribution, and treating it in the same way as any other standard uncertainty associated with an input quantity.

For example, the specification of a particular measurement procedure may require that a defined input quantity be calculated using a specific power series expansion, the higher-order terms of which are known with limited accuracy. A systematic effect arising from the inability to accurately determine these terms results in an unknown constant deviation that cannot be experimentally assessed by repeating the procedure. Therefore, the uncertainty associated with this effect cannot be evaluated or incorporated into the uncertainty of the final measurement result if probability is interpreted strictly in the frequentist sense.

However, interpreting probability as a degree of belief allows the measurement uncertainty associated with this effect to be assessed based on an a priori probability distribution (derived from the available knowledge about the imprecisely known terms) and included in the calculation of the combined standard measurement uncertainty in the same manner as any other uncertainty.

At lower levels of a traceability chain, where it is often assumed that the reference standards used for comparison are known exactly—because they have been calibrated by national metrology institutes against primary standards—the calibration result uncertainty may contain only a Type A standard measurement uncertainty, evaluated from the pooled experimental standard deviation characterizing the measurement.

1.3.2 Evaluation of Type B Measurement Uncertainty

To evaluate the input quantity x_i, which has not been obtained from repeated observations, the associated estimated variance $u^2(x_i)$ or standard measurement uncertainty $u(x_i)$ is determined based on scientific judgment that relies on all available information concerning the possible variability of X_i. Type B standard measurement uncertainty is derived from an assumed probability density function based on the degree of belief that the event will occur (this probability is often referred to as subjective probability).

Sources of information used to estimate Type B measurement uncertainty may include[4,5]:

- known or assumed probability distributions;
- uncertainties associated with physical constants and reference data;
- specifications provided by manufacturers, as well as information from verification, calibration, or other official certificates;
- prior measurement data and general experience with the behavior and properties of relevant materials and instruments.

Correct application of the available information base for the evaluation of Type B standard measurement uncertainty requires sound judgment supported by prior experience and general knowledge. This process is considered a skill that develops with practice. It is important to emphasize that the reliability of a Type B evaluation can be comparable to that of a Type A evaluation, particularly in measurement situations where the latter is based on a small number of statistically independent observations.

When an estimate x_i is obtained from a source such as a manufacturer's specification, a calibration certificate, a reference data set, or similar documentation, and the associated uncertainty is expressed as a multiple of a standard deviation, the standard measurement uncertainty $u(x_i)$ may be determined by dividing the given uncertainty value by the specified multiplier. Accordingly, the estimated variance $u^2(x_i)$ is equal to the square of the resulting standard measurement uncertainty.

For example, a calibration certificate states that the voltage U_s of a reference measuring instrument, with a nominal value of 0.1 V, is 100007 μV, and that the uncertainty associated with this value is 0.1 μV at a coverage factor of three standard deviations. Therefore, the standard measurement uncertainty of the reference voltage standard is: $u(U_s) = (0.1\ \mu\text{V})/3 = 33.33\ \text{nV}$. The estimated variance is: $u^2(U_s) = (33.33\ \text{nV})^2 = 1110.89\ \text{nV}^2$.

The reported uncertainty for a quantity x_i is not always given explicitly as a multiple of a standard deviation. Instead, it may define an interval corresponding to a confidence level of 90%, 95%, 99%, or 99.73%. Unless otherwise specified, it is reasonable to assume that the uncertainty was calculated based on a normal distribution. In such cases, the standard uncertainty for x_i is obtained by dividing the reported uncertainty by the appropriate coverage factor for the normal distribution.

The coverage factors corresponding to the above-mentioned confidence levels are: 1.64 for 90%, 1.96 for 95%, 2.58 for 99%, and 3 for 99.73%.

For example, a calibration certificate states that the resistance R_s of a standard resistor with a nominal value of 100 Ω is 100.000125 Ω ± 235 μΩ at 25°C, and that the quoted uncertainty of 235 μΩ corresponds to a confidence level of 99.73%. The standard measurement uncertainty of the resistor can therefore be taken as: $u(R_s) = (235\ \mu\Omega)/3 = 78.33\ \mu\Omega$. The estimated variance is then: $u^2(R_s) = (78.33\ \mu\Omega)^2 = 6.14 \cdot 10^{-9}\ \Omega^2$.

Let us consider a case where, based on the available information, it can be stated that there is a fifty-fifty chance that the value of the input quantity X_i lies within the interval from α_- to α_+ (in other words, the probability that X_i lies within this interval is 0.5 or 50%). If it can be assumed that the distribution of possible values of X_i is approximately normal, then the best estimate x_i of the quantity X_i can be taken as the midpoint of this interval. Furthermore, if half the width of this interval is denoted by $\alpha = (\alpha_+ - \alpha_-)/2$, then the standard uncertainty can be taken as: $u(x_i) = 1.48\alpha$, because, for a normal distribution with expectation μ_q and standard deviation σ, the interval $\mu_q \pm \sigma/1.48$ includes approximately 50% of the distribution.

If, based on the available information, it can be stated that there is approximately a 2-in-3 chance that the value of X_i lies within the interval from α_- to α_+, then, in other words, the probability that X_i lies within this interval is about 0.67. In this case, it is reasonable to take $u(x_i) = \alpha$, because for a normal distribution with expectation μ_q and standard deviation σ, the interval $\mu_q \pm \sigma$ covers approximately 68.3% of the distribution.

In other cases, only the limits (upper and lower) for X_i can be estimated—for instance, when it can be stated that the probability of X_i falling within the interval from α_- to α_+ is, for all practical purposes, equal to one, and the probability of X_i being outside this interval is zero. If no specific information is available about the possible values of X_i within the interval, it can only be assumed that X_i is equally likely to lie anywhere within it (i.e., a uniform or rectangular distribution of possible values). In this case, the best estimate x_i, the expectation or mean value of X_i, is taken as the midpoint of the interval: $x_i = (\alpha_+ + \alpha_{-,})/2$ with the corresponding variance given by:

$$u^2(x_i) = (\alpha_+ - \alpha_-)^2/12. \tag{1.8}$$

If the difference between the limits $\alpha_+ - \alpha_-$ is denoted as 2α in the case of a symmetric distribution, then from Equation (1.8) we obtain:

$$u^2(x_i) = \alpha^2/3. \tag{1.9}$$

When an uncertainty component obtained in this way makes a significant contribution to the measurement result's overall uncertainty, it makes sense to obtain additional data for its further evaluation.

For example, according to the manufacturer's specifications, for a digital voltmeter used between one and two years after calibration, the maximum permissible error in the 1 V range is defined as the measured value multiplied by $14 \cdot 10^{-6}$, plus the range multiplied by $2 \cdot 10^{-6}$. Suppose the voltmeter is used 20 months after calibration to measure a potential difference V in the 1 V range, and the arithmetic mean of a series of statistically independent repeated observations is $\overline{V} = 0.928571$ V with a Type A standard measurement uncertainty of $u(\overline{V}) = 12\ \mu$V.

Type B measurement uncertainty component associated with the manufacturer's specifications can be evaluated by assuming that the stated error defines a symmetric tolerance interval for an additive correction $\Delta\overline{V}$, with zero expected value, and that any value within the interval is equally probable (i.e., the uniform probability distribution). The half-width α of the uniform distribution can be calculated as: $\alpha = (14 \cdot 10^{-6}) \cdot (0.928571\ \text{V}) + (2 \cdot 10^{-6}) \cdot (1\ \text{V}) = 15\ \mu\text{V}$.

According to Guide to the Expression of Uncertainty in Measurement—GUM (see Equation 1.9), for a uniform distribution with half-width α, the measurement uncertainty is: $u(\Delta\overline{V}) = 8.7\,\mu$V. The best estimate of the measurand V is then: $V = \overline{V} + \Delta\overline{V} = 0.928571$ V. The combined standard measurement uncertainty $u_c(V)$ is calculated by combining the Type A and Type B uncertainty contributions (assuming they are uncorrelated), as: ${u^2}_c(V) = u^2(\overline{V}) + u^2(\Delta\overline{V}) = 2.19 \cdot 10^{-12}\ (V)^2$ and $u_c(\text{V}) = 15\ \mu$V.

When evaluating uncertainty components, it is important to avoid double-counting any individual contribution. If an uncertainty component arising from a specific effect is evaluated as a Type B uncertainty, it should only be included as an independent contribution to the combined standard uncertainty of the measurement result to the extent that the effect does not contribute to the observed variability of repeated observations. This is because the part of the effect that does contribute to the observed variability is already accounted for in the uncertainty component obtained through statistical analysis (Type A evaluation).

The Type B standard measurement uncertainty is evaluated based on a priori information. The process generally involves:

1. based on the mathematical model relating the measurand to the input quantities, a list is compiled that includes:
   i. input quantities that are measured;
   ii. influencing input quantities that are not measured;
   iii. applied corrections for known systematic effects (biases);
   iv. additional sources of error, etc.
2. typically, the evaluation of Type B standard measurement uncertainty is reduced to the application of an appropriate assumed probability distribution.

1.4 FORMS OF EXPRESSION OF UNCERTAINTY COMPONENTS

The measured quantity Y functionally depends on a number of input quantities $X_1, X_2, \ldots, X_N$. These may include both directly measured quantities and quantities that influence the measurement result (such as physical environmental parameters, supply voltage, or external field parameters). This relationship is expressed by the measurement equation, which, in the general case, takes the following form:

$$Y = f(X_1, X_2, \ldots, X_N). \tag{1.10}$$

The input quantities $X_1, X_2, \ldots, X_N$, on which the output quantity Y depends, can themselves be considered as measured quantities. They may also depend on other quantities, including corrections and correction coefficients for systematic effects, resulting in a complex functional relationship f that can never be expressed exactly. Moreover, f can be determined experimentally or may exist only as an algorithm that must be implemented numerically. The function f should be interpreted as a function that includes every quantity, considering all corrections and correction coefficients that may significantly contribute to the measurement result.

Thus, if the data indicate that f does not model the measurement to the extent required by the necessary accuracy of the result, then additional input quantities must be introduced into f to eliminate the inadequacy. This may require introducing an input quantity to represent incomplete knowledge of a phenomenon affecting the measured quantity. However, the measurement Equation (1.10) may be as simple as $Y = X_1 - X_2$. It can represent models such as the comparison of two determinations of the same quantity X.

The set of input quantities $X_1, X_2, \ldots, X_N$ can be divided into the following categories:

- quantities whose values and uncertainties are determined directly in the measurement. These values and uncertainties may be obtained, for example, from a single observation, repeated observations, or an inference based on experience. They may require determining corrections to instrument readings and adjustments for influence quantities such as ambient temperature, atmospheric pressure, and humidity;
- quantities whose values and uncertainties are introduced into the measurement from external sources, such as values related to calibrated standards, certified reference materials, or standard reference data.

The estimate of the measured quantity Y, denoted as y, is obtained from Equation (1.10) using the input estimates $x_1, x_2, \ldots, x_N$ for the N values of the quantities $X_1, X_2, \ldots, X_N$. Thus, the output estimate y, which is the result of the measurement, is expressed as follows:

$$y = f(x_1, x_2, \ldots, x_N). \tag{1.11}$$

In some cases, the estimate y can be obtained from

$$y = \bar{Y} = \frac{1}{N}\sum\nolimits_{k=1}^{N} Y_k = \frac{1}{N}\sum\nolimits_{N=1}^{k} f(X_{1,k}, X_{2,k}, \ldots, X_{N,k}). \tag{1.12}$$

Thus, y is taken as the arithmetic mean or the average value of N independent determinations Y_k of the quantity Y, where each determination has its own uncertainty and is based on a complete set of observed values of the N input quantities X_i, obtained at the same time. This method of averaging—instead of using

$y = f(X_1, X_2, \ldots, X_N)$, where each X_i is the arithmetic mean of the individual observations $\bar{X}_i = \frac{1}{N}\sum_{k=1}^{N} X_{i,k}$ — may be preferable when f is a nonlinear function of the input quantities $X_1, X_2, \ldots, X_N$. However, the two approaches are equivalent if f is a linear function.

1.4.1 Form of Expression of Standard Measurement Uncertainty

The input data for determining Type A standard measurement uncertainty are the results of repeated measurements. The Type A standard measurement uncertainty of a single measurement is determined using formula (1.2), while the Type A standard measurement uncertainty of the mean value is calculated using formula (1.4).

Type B standard measurement uncertainty components are generally determined based on information about the lower and upper bounds $[\alpha_-; \alpha_+]$ of a presumed (a priori defined) probability distribution, or through an interval U associated with a given confidence level p.

To determine the Type B standard measurement uncertainty, one must take the positive square root of the sum of the products of the confidence level of each value and the square of its deviation. Thus, the general form of the formula for calculating Type B standard measurement uncertainty for discrete data is as follows:

$$u_B(X) = \sqrt{\sum_{i=1}^{n}\left(x_i - \sum_{i=1}^{n} x_i p_i\right)^2 p_i} = \sqrt{\sum_{i=1}^{n}(x_i - \bar{x})^2 p_i}. \quad (1.13)$$

The Type B standard measurement uncertainty for a continuous input quantity X is evaluated using the formula:

$$u_B(X) = \sqrt{\int_{-\infty}^{+\infty}\left(x - \int_{-\infty}^{+\infty} xp(x)dx\right)^2 p(x)dx} = \sqrt{\int_{-\infty}^{+\infty}(x - \bar{x})^2 p(x)dx}. \quad (1.14)$$

If the lower and upper bounds $[\alpha_-; \alpha_+]$ for the value of the quantity X_i can be estimated, then the Type B standard measurement uncertainties can be determined—under an assumption about the possible form of the probability distribution—using formulas:

1. For a triangular distribution

$$u_B(X_i) = \frac{\alpha_+ - \alpha_-}{\sqrt{24}}. \quad (1.15)$$

2. For a trapezoidal distribution

$$u_B(X_i) = \frac{[\alpha_+ - \alpha_-]\sqrt{1+\beta^2}}{\sqrt{24}}, \quad (1.16)$$

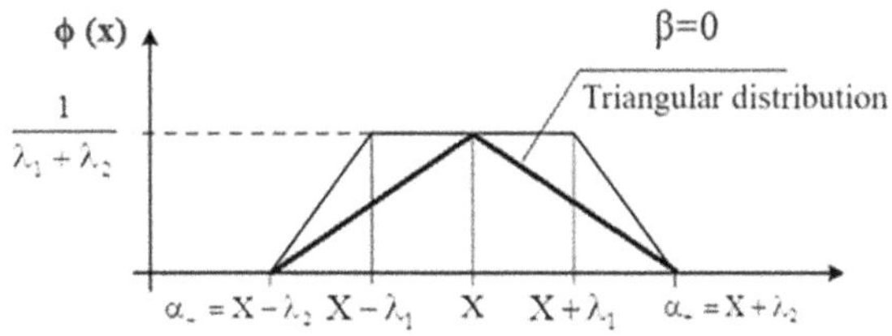

FIGURE 1.3 Trapezoidal distribution.

where β is a parameter defined by the following ratio:

$$\beta = \frac{\lambda_1}{\lambda_2}, \tag{1.17}$$

where $\lambda_1 = \frac{|(\alpha_{1+} - \alpha_{1-}) - (\alpha_{2+} - \alpha_{2-})|}{2}$, $\lambda_2 = \frac{\alpha_+ - \alpha_-}{2}$, $\alpha_+ = \alpha_{1+} + \alpha_{2+}$, $\alpha_- = \alpha_{1-} + \alpha_{2-}$ (see Figure 1.3).

As β varies from 0 to 1, the trapezoidal distribution changes from triangular ($\beta = 0$) to uniform ($\beta = 1$).

3. For a uniform distribution (see Figure 1.4)

$$u_B(X_i) = \frac{\alpha_+ - \alpha_-}{\sqrt{12}}. \tag{1.18}$$

4. For an exponential distribution

$$u_B(X_i) = \sqrt{\frac{(\alpha_+ - x)(x - \alpha_-) - (\alpha_+ - 2x + \alpha_-)}{\lambda}}, \tag{1.19}$$

where x is the expected value; λ is the parameter of the exponential distribution (see Figure 1.5).

5. For an arcsine distribution (see Figure 1.6)

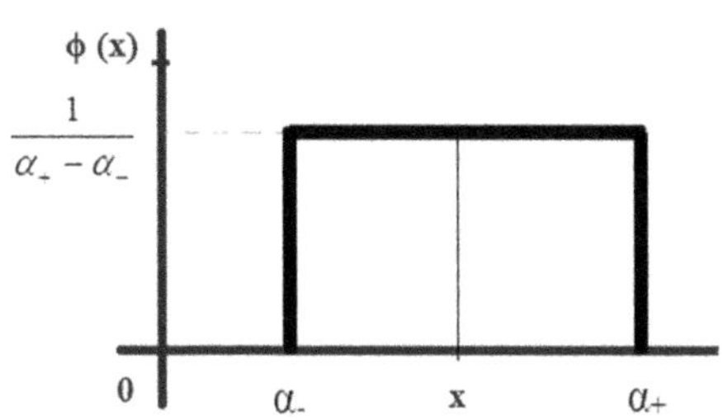

FIGURE 1.4 Uniform distribution.

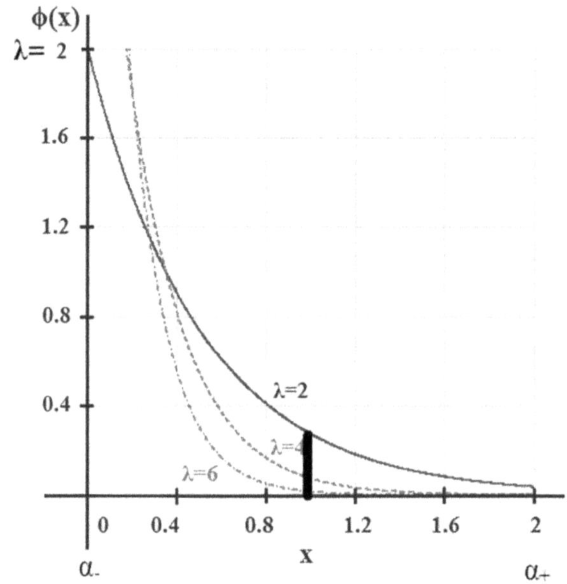

FIGURE 1.5 Exponential distribution.

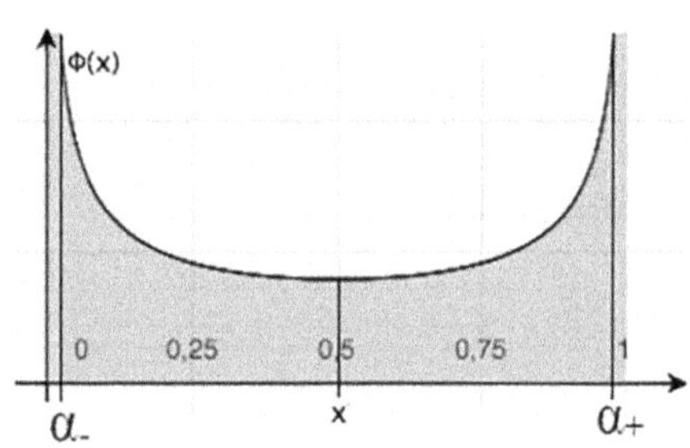

FIGURE 1.6 Arcsine distribution.

$$u_B(X_i) = \frac{\alpha_+ - \alpha_-}{\sqrt{8}}. \tag{1.20}$$

6. For given intervals U_p with a known confidence level p, assuming a normal distribution, the Type B uncertainty is determined using the following formula:

$$u_B(X_i) = \frac{U_p}{k_p}, \tag{1.21}$$

where k_p is the coverage factor, which for a normal distribution equals 1.64, 1.96, 2.58, and 3 for confidence levels of 0.9, 0.95, 0.99, and 0.9973, respectively.

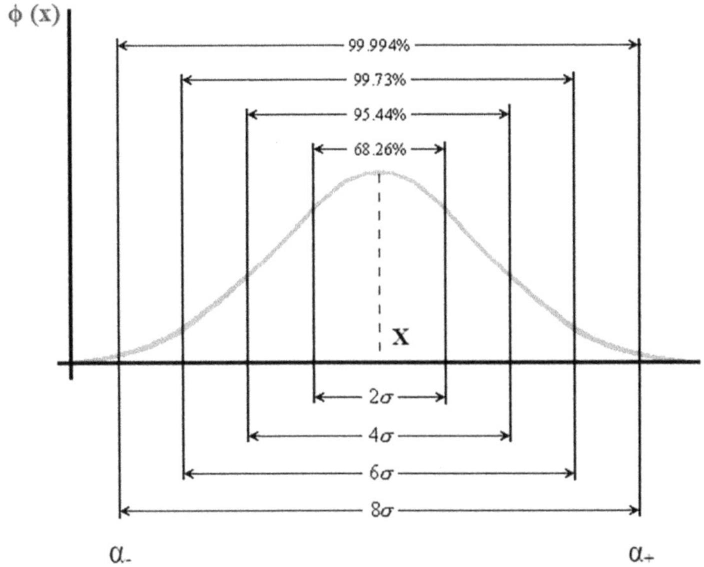

FIGURE 1.7 Normal (Gaussian) distribution.

For example, for a measurement result described by a normal distribution, the Type B standard uncertainty at a confidence level of $p = 99.73\%$ (see Figure 1.7) can be estimated using the following formula:

$$u_B(X_i) = \frac{\alpha_+ - \alpha_-}{6}. \tag{1.22}$$

For a multivariate Gaussian distribution with an N-dimensional quantity $\boldsymbol{X} = (X_1, X_2, \ldots, X_N)^T$, having the best estimate $\boldsymbol{x} = (x_1, x_2, \ldots, x_N)^T$, the positive definite uncertainty matrix U_x is given as follows:

$$U_x = \begin{bmatrix} u^2(x_1) & u(x_1, x_2) & \ldots & u(x_1, x_N) \\ u(x_2, x_1) & u^2(x_2) & \ldots & u(x_2, x_N) \\ \cdot & \cdot & \cdot & \cdot \\ \cdot & \cdot & \cdot & \cdot \\ \cdot & \cdot & \cdot & \cdot \\ u(x_N, x_1) & u(x_N, x_2) & \ldots & u^2(x_N) \end{bmatrix}. \tag{1.23}$$

The quantity $\boldsymbol{X}$ may be assigned a multivariate Gaussian distribution $N(\boldsymbol{x}, U_x)$.

The covariance (or variance–covariance) matrix for the vector random variable $\boldsymbol{X}$ is described by the following expression:

$$V(\boldsymbol{X}) = U_x. \tag{1.24}$$

If the elements of the vector $\boldsymbol{X}$ are uncorrelated, then $u(x_i, x_j) = 0$.

The covariance matrix $V(\boldsymbol{X})$, given known intervals $\boldsymbol{U_p}$ with known confidence levels $\boldsymbol{p}$, is determined using the following expression:

$$\boldsymbol{U_p} = \boldsymbol{R}^T \boldsymbol{k_p} = U_x \boldsymbol{R}^{-1} \boldsymbol{k_p}, \tag{1.25}$$

where $\boldsymbol{k_p} = (k_{p1}, k_{p2}, \ldots, k_{pN})^T$ is the vector of coverage factors for the given confidence levels $p1, p2, \ldots, p_N$; $\boldsymbol{R}$ is an upper triangular matrix obtained from the Cholesky decomposition $U_x = \boldsymbol{R}^T \boldsymbol{R}$.

Instead of the Cholesky decomposition $U_x = \boldsymbol{R}^T \boldsymbol{R}$, any matrix factorization of this form can be used.

Thus, for known (given) intervals $\boldsymbol{U_p}$, assuming an N-dimensional Gaussian distribution, the Type B uncertainty matrix associated with the vector estimate of the input quantity $\boldsymbol{x}$ is determined by the following formula:

$$U_x = \boldsymbol{U_p R k_p}^{-1}. \tag{1.26}$$

If the covariance matrix $V(\boldsymbol{X})$, described by formula (1.24), is positive definite—that is, all its eigenvalues are strictly positive—then the Cholesky factor $\boldsymbol{R}$ is unique.

If the covariance matrix $V(\boldsymbol{X})$ is not positive definite, for example, due to rounding errors in numerical methods or other sources, then $\boldsymbol{R}$ may not exist. In such cases, it is recommended to "adjust" the values of $V(\boldsymbol{X})$, that is, to make the smallest possible modifications (as much as feasible) to $V(\boldsymbol{X})$ so that the Cholesky factor $\boldsymbol{R}$ of the modified matrix is positive definite.

If the covariance matrix $V(\boldsymbol{X})$ is positive definite, then it is also possible to form a decomposition of the form:

$$V(\boldsymbol{X}) = U_x = \boldsymbol{Q D Q}^T, \tag{1.27}$$

where $\boldsymbol{Q}$ is an orthogonal matrix; $\boldsymbol{D}$ is a diagonal matrix.

In this case, the product $\boldsymbol{D Q}^T$ can be used to generate random values from $N(0, V(\boldsymbol{X}))$ when modeling standard normally distributed random variables, as well as in numerical methods and software for uncertainty calculations, even if the rank of $V(\boldsymbol{X})$ is insufficient.

The measurement uncertainty of constants, coefficients, and corrections to constants and coefficients, where the dispersion intervals correspond to the least significant digit of their numerical values, is calculated using the formula under the assumption of a uniform distribution of these values:

$$u_B = \frac{q}{\sqrt{12}}, \tag{1.28}$$

where q is the least significant digit of the numerical value.

For symmetrical limits $\pm \alpha_i$, the Type B standard measurement uncertainty, in the absence of information about the probability distribution of the input quantity X_i, is determined using the following formula[6]:

$$u_B(X_i) = \frac{2\alpha_i}{\sqrt{12}} = \frac{\alpha_i}{\sqrt{3}}. \qquad (1.29)$$

The equation for determining the standard measurement uncertainty of reading from an analog scale of a measuring instrument, under the assumption of a uniform distribution, is given by

$$u_B(X_i) = \frac{\left(x + \frac{C}{4}\right) - \left(x - \frac{C}{4}\right)}{\sqrt{12}} = \frac{C}{4\sqrt{3}}, \qquad (1.30)$$

where x is the measured value of the quantity; C is the scale division value of the measuring instrument.

If the scale is non-uniform, the standard measurement uncertainty is determined separately for each range with a defined scale division value.

For a measuring instrument with a specified accuracy class, the standard uncertainty is calculated using the following formula:

$$u_B = \frac{\Delta_{MI}}{\sqrt{3}}, \qquad (1.31)$$

where Δ_{MI} is the limit of the maximum permissible (instrumental) error, which is determined by the formula:

$$\Delta_{MI} = \pm \frac{\gamma \cdot X_R}{100\%}, \qquad (1.32)$$

where γ is the accuracy class of the measuring instrument (expressed as a percentage); R is the rated value of the instrument's measurement range, which depends on the type of its scale (if the measuring instrument's scale has a display range from zero to a maximum value $X_{\max}$, the rated value X_R is taken to be the maximum value of the instrument's display range $X_R = X_{\max}$).

Thus, the calculation of Type B standard measurement uncertainty is possible when certain information about the quantity being evaluated is available. Such information includes knowledge of the probability distribution and its bounds. These can be obtained from:

1. previous (archival) or special additional measurements;
2. manufacturer's specifications (documentation) for the measuring instruments and materials used during the measurement process;
3. calibration and verification certificates or protocols for the measuring instruments;
4. reference data;
5. knowledge of the behavior and properties of substances, materials, or instruments.

A common situation occurs when the distribution limits are known, but no information about the probability distribution within those limits is available. In such cases, the calculation of Type B standard uncertainty is based on the formula for a uniform distribution.

1.4.2 Form of Expression of Combined Uncertainty for Uncorrelated Input Quantities

The standard uncertainty of the measured quantity Y estimate, and consequently the measurement result, is formed by appropriately combining the standard uncertainties of the input estimates $x_1, x_2, \ldots, x_N$. This standard measurement uncertainty of the estimate y is denoted as $u_c(y)$ and is called the combined standard uncertainty.

Each input estimate x_i and its associated standard measurement uncertainty $u_c(x_i)$ are derived from the probability distribution of possible values of the input quantity X_i. As previously mentioned, this probability distribution may be based on observed data sets $X_{i,k}$ of the quantity X_i, or it may be an a priori distribution.

Measurement results are considered uncorrelated when all input quantities are independent.

The combined measurement uncertainty $u_c(y)$ is the positive square root of the combined variance $u^2{}_c(y)$, which is calculated by the formula:

$$u_c^2(y) = \sum\nolimits_{i=1}^{N} \left(\frac{\partial f}{\partial x_i}\right)^2 u^2(x_i), \tag{1.33}$$

where f is the function given in Equation (1.11); $u(x_i)$ is the standard measurement uncertainty, evaluated as Type A or Type B, as described earlier.

The combined measurement uncertainty $u_c(y)$ is the estimated standard deviation and characterizes the dispersion of values that can reasonably be attributed to the measured quantity Y.

Equation (1.33) is obtained by approximating the measurement Equation (1.11) using a first-order Taylor series expansion and represents the law of propagation of uncertainty.

In cases of significant nonlinearity of the model equation f, higher-order terms of the Taylor series expansion must be included in Equation (1.33) to properly determine the combined variance (i.e., the uncertainty squared) $u^2{}_c(y)$.

When the distribution of each X_i is symmetric about its mean value, these higher-order terms become significant and must be added to the terms in Equation (1.33):

$$u_c^2(y) = \sum_{i=1}^{N} \left(\frac{\partial f}{\partial x_i}\right)^2 u^2(x_i) + \sum\nolimits_{i=1}^{N} \sum\nolimits_{j=1}^{N} \left[\frac{1}{2}\left(\frac{\partial^2 f}{\partial x_i \partial x_j}\right)^2 + \frac{\partial f}{\partial x_i} \frac{\partial^3 f}{\partial x_i \partial x_j^2}\right] u^2(x_i) u^2(x_j). \tag{1.34}$$

Equation (1.34) can be used to estimate the combined standard uncertainty only when the input quantities are uncorrelated.

In this case, the mean value $\bar{Y}$ is determined using the following equation:

$$\bar{Y} \approx f(\bar{x}_1, \bar{x}_2, \ldots, \bar{x}_N) + \frac{1}{2} \sum\nolimits_{i=1}^{N} \left(\frac{\partial f}{\partial x_i}\right)^2 u^2(x_i) + \sum\nolimits_{i<j}^{N} \frac{\partial^2 f}{\partial x_i \partial x_j} u(x_i, x_j). \tag{1.35}$$

In indirect measurements, when the random input quantities are independent, the combined variance (i.e., the square of the combined standard uncertainty) is determined using the following equation:

$$u_c^2(y) = \sum_{i=1}^{N} \left(\frac{\partial f}{\partial x_i}\right)^2 u^2(x_i) + \frac{1}{4}\sum_{i=1}^{N} \left(\frac{\partial^2 f}{\partial x_i^2}\right)^2 \left[\mu_4(x_i) - u^4(x_i)\right] +$$

$$+\sum_{i<j}^{N} \frac{\partial^2 f}{\partial x_i \partial x_j} u^2(x_i) u^2(x_j) + \sum_{i=1}^{N} \left(\frac{\partial f}{\partial x_i}\right)\left(\frac{\partial^2 f}{\partial x_i^2}\right) \mu_3(x_i), \qquad (1.36)$$

where $\mu_3(x_i)$ and $\mu_4(x_i)$ are the third and fourth central moments of the distribution of the input quantities, respectively.

From Equation (1.36), it is evident that when calculating the combined uncertainty under conditions of significant nonlinearity of the transformation function (model equation), it is necessary to account for the shape and asymmetry of the input distributions. For a normal distribution, the results obtained using Equations (1.34) and (1.36) coincide. However, it should be noted that Equation (1.34) is valid only under the assumption that the input quantities of the measurement equation Y follow a normal distribution.

The partial derivatives $\frac{\partial f}{\partial x_i}$ in expression (1.33) are equal to $\frac{\partial f}{\partial X_i}$ evaluated at $X_i = x_i$. These derivatives are called sensitivity coefficients and indicate how the output estimate y changes in response to changes in the input estimates $x_1, x_2, \ldots, x_N$.

In particular, the change in y caused by a small change Δx_i in the input estimate x_i is determined by the equation $(\Delta y)_i = \frac{\partial f}{\partial x_i} \cdot (\Delta x_i)$. If this change is due to the standard uncertainty of the estimate x_i, then the corresponding change in y is $\frac{\partial f}{\partial x_i} \cdot u(x_i)$.

Therefore, the total variance $u_c{}^2(y)$ can be viewed as the sum of terms, each of which represents the estimated variance associated with the output estimate y, caused by the estimated variance associated with each input estimate x_i. This leads to expressing Equation (1.33) in the following form:

$$u_c{}^2(y) = \sum_{i=1}^{N} [c_i \cdot u(x_i)]^2 = \sum_{i=1}^{N} u_i^2(y), \qquad (1.37)$$

where $c_i = \frac{\partial f}{\partial x_i}, u_c(y) = \mid c_i \mid u(x_i)$.

The partial derivatives $\frac{\partial f}{\partial x_i} = \frac{\partial f}{\partial X_i}$ are ideally evaluated at the expected values of X_i. However, in practice, they are typically evaluated at the values $\frac{\partial f}{\partial x_i} = \frac{\partial f}{\partial X_i} \mid x_1, x_2, \ldots, x_N$. The sensitivity coefficients $\frac{\partial f}{\partial x_i}$, instead of being calculated analytically from the function f, are sometimes determined experimentally by measuring the change in Y caused by a change in the selected input quantity X_i, while keeping all other input quantities constant. In this case, knowledge of the function f is reduced to an empirical first-order Taylor expansion based on the measured sensitivity coefficients.

If Y is of the form $Y = c \cdot X_1{}^{p1} \cdot X_2{}^{p2} \cdot \ldots \cdot X_N{}^{p^N}$, and it is known that the exponents p_i are positive or negative numbers with small uncertainties, then the total variance given by Equation (1.33) can be expressed as:

$$\left[\frac{u_c(y)}{y}\right]^2 = \sum_{i=1}^{N} \left[\frac{p_i u(x_i)}{x_i}\right]^2. \tag{1.38}$$

Equation (1.38) has the same form as Equation (1.37), but with the combined variance ${u_c}^2(y)$ expressed as the relative combined variance $\left[\frac{u_c(y)}{y}\right]^2$, and the estimated variances ${u_c}^2(x_i)$ associated with each input estimate expressed as estimated relative variances $\left[\frac{u(x_i)}{x_i}\right]^2$.

Combined uncertainties for different types of measurement equations are summarized in Table 1.1.

TABLE 1.1
Combined Standard Uncertainties for Various Forms of Measurement Equations

Measurement Equation	Combined Standard Uncertainties
$Y = X_1 \pm X_2 \pm \ldots \pm X_N$	$u_c(y) = \left[\sum_{i=1}^{N} u^2(x_i)\right]^{0.5}$
$Y = C_1X_1 \pm C_2X_2 \pm \ldots \pm C_NX_N$	$u_c(y) = \left[\sum_{i=1}^{N} C_i^2 u^2(x_i)\right]^{0.5}$
$Y = \prod_{i=1}^{N} X_i = X_1 \cdot X_2 \cdot \ldots \cdot X_N$ $Y = \frac{X_1 \cdot X_2 \cdot X_3 \cdot X_4}{X_5 \cdot X_6 \cdot X_7 \cdot X_8}$ $Y = X_1/X_2/X_3/\ldots/X_N$	$\tilde{u}_c(y) = \sqrt{\sum_{i=1}^{N} \frac{u^2(x_i)}{x_i^2}}$, where $\tilde{u}_c(y) = \frac{u_c(y)}{y}$, $\tilde{u}(x_i) = \frac{u(x_i)}{x_i}$
$Y = X_1 \cdot X_2 \cdot X_3 + X_4 \cdot X_5 \cdot X_6 + \ldots + X_{N-2} \cdot X_{N-1} \cdot X_N$	$u_c^2(y) = (x_1 \cdot x_2 \cdot x_3)^2 \left(\tilde{u}^2(x_1) + \tilde{u}^2(x_2) + \tilde{u}^2(x_3)\right) + (x_4 \cdot x_5 \cdot x_6)^2 \left(\tilde{u}^2(x_4) + \tilde{u}^2(x_5) + \tilde{u}^2(x_6)\right) + \ldots + (x_{N-2} \cdot x_{N-1} \cdot x_N)^2 \left(\tilde{u}^2(x_{N-2}) + \tilde{u}^2(x_{N-1}) + \tilde{u}^2(x_N)\right)$, where $\tilde{u}(x_N) = u(x_N)/x_N$
$Y = X^m$	$u_c(y) = m \cdot x^{m-1} u(x)$, $\tilde{u}_c(y) = u_c(y)/y = m \cdot u(x)/x$
$Y = \ln(X_1/X_2) = \ln X_1 - \ln X_2$ $Y = \ln(X_1X_2) = \ln X_1 + \ln X_2$	$\tilde{u}_c(y) = \left[u^2(x_1)/(x_1^2) + u^2(x_2)/(x_2^2)\right]^{0.5}$
$Y = (X_1 \pm X_2)(X_3 \pm X_4)$ $Y = (X_1 \pm X_2)/(X_3 \pm X_4)$	$\tilde{u}_c(y) = \left[\frac{u^2(x_1) + u^2(x_2)}{(x_1 + x_2)^2} + \frac{u^2(x_3) + u^2(x_4)}{(x_3 + x_4)^2}\right]^{0.5}$
$Y = {X_1}^m \pm X_2^z$	$u_c(y) = \left[\left[m \cdot x_1^{m-1} u(x_1)\right]^2 + \left[z \cdot {x_2}^{z-1} u(x_2)\right]^2\right]^{0.5}$
$Y = {X_1}^m \cdot X_2^z$ $Y = {X_1}^m / X_2^z$	$\tilde{u}_c(y) = \left[\left(m \frac{u(x_1)}{x_1}\right)^2 + \left(z \frac{u(x_2)}{x_2}\right)^2\right]^{0.5}$
$Y == ({X_1}^m \pm {X_2}^z)/({X_3}^k \pm {X_4}^p)$ $Y = ({X_1}^m \pm {X_2}^z) \cdot ({X_3}^k \pm {X_4}^p)$	$\tilde{u}_c^2(y) = \frac{\left(mx_1^{m-1} u(x_1)\right)^2 + \left(zx_2^{z-1} u(x_2)\right)^2}{\left(x_1^m + {x_2}^z\right)^2} + \frac{\left(kx_3^{k-1} u(x_3)\right)^2 + \left(px_4^{p-1} u(x_4)\right)^2}{\left(x_3^k + {x_4}^p\right)^2}$

1.4.3 Form of Expression of Combined Uncertainty for Correlated Input Quantities

Equation (1.33) and the equations derived from it, such as Equations (1.34) and (1.35)–(1.38), are valid only if the input quantities X_i are independent or uncorrelated. If any of the X_i are significantly correlated, this correlation must be considered.

When the input quantities are correlated, the expression for the combined variance $u_c{}^2(y)$ associated with the measurement result takes the following form:

$$u_c{}^2(y) = \sum_{i=1}^{N}\sum_{j=1}^{N}\frac{\partial f}{\partial x_i}\frac{\partial f}{\partial x_j}u(x_i,x_j) =$$

$$= \sum_{i=1}^{N}\left(\frac{\partial f}{\partial x_i}\right)^2 u^2(x_i) + 2\sum_{i=1}^{N-1}\sum_{j=i+1}^{N}\frac{\partial f}{\partial x_i}\frac{\partial f}{\partial x_j}u(x_i,x_j), \tag{1.39}$$

where x_i and x_j are the estimates of X_i and X_j; $u(x_i,x_j) = u(x_j,x_i)$ is the estimated covariance between x_i and x_j.

Covariance is a property of a pair of random variables, and for two continuous random variables X_1 and X_2 characterized by a joint probability density function $p(x)$, it is expressed by the following equation:

$$u(x_1,x_2) = \int_{-\infty}^{+\infty}\int_{-\infty}^{+\infty}(x_1-\bar{x}_1)(x_2-\bar{x}_2)p(x_1,x_2)dx_1dx_2. \tag{1.40}$$

For discrete, equally accurate input measured quantities, the covariance is calculated using the equation:

$$u(x_1,x_2) = \frac{1}{N}\sum_{l=1}^{N_{ij}}(x_{i_l}-\bar{x}_i)(x_{j_l}-\bar{x}_j). \tag{1.41}$$

The degree of correlation between x_i and x_j is characterized by the estimated correlation coefficient[7]:

$$r(x_i,x_j) = \frac{u(x_i,x_j)}{u(x_i)u(x_j)} = \frac{\sum_{l=1}^{N_{ij}}(x_{i_l}-\bar{x}_i)(x_{j_l}-\bar{x}_j)}{\sqrt{\sum_{l=1}^{N_{ij}}(x_{i_l}-\bar{x}_i)^2\sum_{l=1}^{N_{ij}}(x_{j_l}-\bar{x}_j)^2}}, \tag{1.42}$$

where $r(x_i,x_j) = r(x_j,x_i)$ is the correlation coefficient, which can range between $-1 < r(x_i,x_j) < +1$. If the estimates x_i and x_j are independent, then $r(x_i,x_j) = 0$, and a change in one does not imply an expected change in the other.

In terms of correlation coefficients, which are generally easier to interpret than covariances, the expression for the combined variance (1.39) can be written as:

$$u_c^2(y) = \sum_{i=1}^{N}c_i^2u^2(x_i) + 2\sum_{i=1}^{N-1}\sum_{j=i+1}^{N}c_ic_ju(x_i)u(x_j)r(x_i,x_j). \tag{1.43}$$

In the case where all input estimates are correlated with correlation coefficients $r(x_i, x_j) = 1$, Equation (1.43) reduces to the form:

$$u_c^2(y) = \sum_{i=1}^{N} [c_i u(x_i)]^2 = \left[\sum_{i=1}^{N} \frac{\partial f}{\partial x_i} u(x_i)\right]^2. \tag{1.44}$$

Thus, the combined standard uncertainty $u_c(y)$ is the positive square root of the linear sum of terms representing the contribution of the standard uncertainty of each input estimate x_i to the variance of the output estimate y. This linear sum should not be confused with the general law of propagation of errors, even though they may appear similar in form; standard uncertainties are not errors.

If the transformation function Y is a polynomial of the form $Y = cX_1^{p1} \cdot X_2^{p2} \cdot \ldots \cdot X_N^{pN}$, and the input quantities X_i are correlated, then the right-hand side of Equation (1.38) must be supplemented with the term:

$$+2\sum_{i=1}^{N-1} \sum_{j=i+1}^{N} \left[\frac{p_i u(x_i)}{x_i}\right] \left[\frac{p_j u(x_j)}{x_j}\right] r(x_i, x_j). \tag{1.45}$$

Considering the specifics of measurement uncertainty, correlations between input quantities must not be ignored if they exist and are significant. The associated covariances should be evaluated either experimentally, by varying the correlated input quantities (Type A evaluation), or by using all available information about the correlated variability of the quantities (Type B evaluation). Sound judgment, based on prior experimental results and general knowledge, is especially important when estimating the degree of correlation between input quantities caused by common influencing factors such as ambient temperature, atmospheric pressure, and humidity.

Fortunately, in many practical cases, the effects of such influences are only weakly correlated, making it reasonable to assume that the input quantities affected by them are uncorrelated. However, if it cannot be assumed that they are uncorrelated, the correlations themselves can be removed by explicitly introducing the common influencing factors into the measurement model as additional independent input quantities.

A correlation between two input quantities may also arise when the same measuring instrument, physical standard, or reference data—subject to significant uncertainty—is used in their determination. For example, if a temperature correction required to estimate input quantity X_i is obtained using a particular thermometer, and the same correction is also used to estimate input quantity X_j, then X_i and X_j may be correlated. However, if X_i and X_j are determined as uncorrected quantities, or if the parameters defining the calibration curve of the thermometer are included in the measurement model as additional independent input quantities with their own standard uncertainties, the correlation between X_i and X_j is effectively removed.

1.4.4 Criterion for Testing the Presence of Correlation between Pairs of Measurement Results

When evaluating and presenting the uncertainties of measurement results, it is first necessary to determine whether a correlation exists between any pairs of indirect measurement results. To this end, the following criterion is proposed for testing the presence of a correlation between measurement results:

$$\frac{r(x_i, x_j)\sqrt{N}}{1 - r^2(x_i, x_j)} > t_p(\nu_{\text{eff}}), \tag{1.46}$$

where $t_p(\nu_{\text{eff}})$ is the quantile of Student's t-distribution with an effective number of degrees of freedom ν_{eff} and level of confidence p; N is the number of matched pairs of measurement results; $r(x_i, x_j)$ is the correlation coefficient between the measurement pairs x_i and x_j, calculated using Equation (1.42).

If inequality (1.46) is satisfied, it indicates that a correlation exists between the given pair of measurement results x_i and x_j, and the combined uncertainty must be calculated using Equation (1.39) or (1.43).

However, if inequality (1.46) is not satisfied, this indicates that the correlation is either absent or insignificant, and the combined uncertainty should be calculated using Equation (1.33). In cases where the transformation function (model equation) is highly nonlinear, Equation (1.34) or (1.36) should be used instead.

1.4.5 Form of Expression of Expanded Measurement Uncertainty

The international standard ISO/IEC GUIDE 98-3:2008[2] and the Joint Committee for Guides in Metrology (JCGM) Recommendation 100:2008[1] endorse the use of combined standard uncertainty $u_c(y)$ as a parameter for quantitatively expressing the measurement uncertainty of a result.

Although the combined standard uncertainty $u_c(y)$ can be widely used to express the uncertainty of a measurement result, in certain cases—such as in trade, industry, regulatory contexts, and matters concerning health and safety—it is advisable to additionally specify an interval measure of uncertainty that defines an interval within which the true value of the measurand is asserted to lie. The necessity of such a requirement has been recognized by the international standard ISO/IEC GUIDE 98-3:2008, the JCGM Recommendation 100:2008, and National Institute of Standards and Technology (NIST).

This additional measure of uncertainty, corresponding to an interval estimate, is called the expanded uncertainty and is denoted by the symbol U. The expanded uncertainty is obtained by multiplying the combined standard uncertainty $u_c(y)$ by a coverage factor k:

$$U = k \cdot u_c(y). \tag{1.47}$$

The measurement result is expressed as $Y = y \pm U$, which means that the best estimate of the value attributed to the quantity Y is y, and that the interval from $y - U$ to $y + U$ contains a large portion of the distribution of values that can reasonably be attributed to the measurand Y. This interval can also be represented as: $y - U \leq Y \leq y + U$.

The terms "confidence interval" and "confidence factor" have specific definitions in statistics and apply to the interval defined by U only under certain conditions, including the condition that all uncertainty components contributing to $u_c(y)$ were obtained by Type A evaluation. In uncertainty theory, when considering U as an interval around the measurement result that contains a large part of the probability distribution p, it is referred to as the coverage probability or confidence factor for that interval.

The value of the coverage factor k is chosen based on the desired level of confidence for the interval from $y - U$ to $y + U$. Typically, the coverage factor k ranges from 2 to 3. However, in special cases, k may fall outside this range. Extensive experience and a thorough understanding of how the measurement result is applied can facilitate

$$U = k \cdot u_c(y) = t_p(v) \cdot u_c(y), \tag{1.48}$$

where $t_p(v)$ is Student's t-distribution coefficient for coverage probability p and degrees of freedom $v = N - 1$.

The degrees of freedom $v = N - 1$ for a single quantity estimated as the arithmetic mean of N independent observations.

As $v \rightarrow \infty$, the t-distribution approaches the normal distribution, and approximate values of Student's t-coefficient can be calculated using the equation:

$$t_p(v) \approx k\,(1 + 2/v)^{1/2}, \tag{1.49}$$

where k is the coverage factor required to obtain an interval with level of confidence p for a normal distribution.

To obtain a more accurate approximation for estimating the expanded uncertainty, the t-distribution should be used. However, in the general case, the t-distribution will not describe $(y - Y)/u_c(y)$ if $u_c{}^2(y)$ is the sum of two or more evaluated variance components $u_i(y) = c_i{}^2 \cdot u^2(x_i)$, even if each x_i is an estimate of a normally distributed input quantity X_i. Nevertheless, the distribution of this variable can be approximated by a t-distribution with an effective degrees of freedom v_{eff}, obtained from the Welch–Satterthwaite equation:

$$v_{\text{eff}} = \frac{u_c^4(y)}{\sum_{i=1}^{N} \frac{u_i^4(y)}{v_i}}. \tag{1.50}$$

When processing standard uncertainties of Type A and Type B separately, it is necessary to additionally calculate and report the values of v_{effA} and v_{effB}, computed using Equation (1.50).

When summing the uncertainties of mean values of input quantities determined by Type A evaluation, the degrees of freedom v should be chosen as $N - 1$.

When summing uncertainties of input quantities determined by Type B evaluation, the degrees of freedom v is assumed to be infinite (∞).

If the combined standard uncertainty $u_c(y)$ is evaluated separately by Type A $u_{c_A}(y)$ and Type B $u_{c_B}(y)$ methods, then the mentioned quantities are related by the following relationships $u_c^2(y) = u_{c_A}^2(y) + u_{c_B}^2(y)$. Considering Equation (1.50) can be written in the following form:

$$\nu_{\text{eff}} = (N-1)\frac{\left[u_{c_A}^2(y) + u_{c_B}^2(y)\right]^2}{u_A^4(y)}. \quad (1.51)$$

1.4.6 Form of Expression of Relative Measurement Uncertainties

Considering that a relative quantity is the ratio of two homogeneous physical quantities, the relative uncertainty is defined as the ratio of the standard, combined, or expanded uncertainty to the estimate of the measured quantity. The expressions for determining the relative standard, relative combined, and relative expanded uncertainties are as follows:

$$\tilde{u}_{A(B)} = \frac{u_{A(B)}(x)}{|x|}, \text{ for } |x| \neq 0; \quad (1.52)$$

$$\tilde{u}_c = \frac{u_c(y)}{|y|}, \text{ for } |y| \neq 0; \quad (1.53)$$

$$\tilde{U} = \frac{U}{|y|}, \text{ for } |y| \neq 0. \quad (1.54)$$

The relative uncertainties mentioned above can be expressed not only as relative values but also as percentages. To do this, Equations (1.52)–(1.54) should be multiplied by 100%.

Notes

1 JCGM 100:2008, *Evaluation of measurement data—Guide to the expression of uncertainty in measurement* (Int. Organ. Stand., Geneva, ISBN 50, 2008: 134).

2 ISO/IEC *Guide 98-1:2024, Guide to the expression of uncertainty in measurement* (Int. Organ. Stand., Geneva, 2024: 13).

3 ISO/IEC Guide 98-3:2008, *Uncertainty of measurement Part 3: Guide to the expression of uncertainty in measurement* (GUM:1995) (Int. Organ. Stand., Geneva, 2008: 120).

4 JCGM 101:2008, *Evaluation of measurement data – Supplement 1 to the "Guide to the expression of uncertainty in measurement" – Propagation of distributions using a Monte Carlo method* (Int. Organ. Stand., Geneva, 2008: 82).

5 JCGM 102:2011, *Evaluation of measurement data – Supplement 2 to the "Guide to the expression of uncertainty in measurement" – Extension to any number of output quantities* (Int. Organ. Stand. Geneva, 2011: 72).

6 B. N. Taylor and C. E. Kuyatt, *Guidelines for evaluating and expressing the uncertainty of NIST measurement results*, NIST Technical Note 1297, Rev. ed., National Institute of Standards and Technology, Gaithersburg, MD, 2008. [Online]. Available: https://doi.org/10.6028/NIST.TN.1297.

7 O. M. Vasilevskyi, V. Yu. Kucheruk, Ye. T. Volodarskyi, *Uncertainty of measurement, control and test results* (OLDI-PLUS, 2020: 352).

BIBLIOGRAPHY

ISO/IEC Guide 98-1:2024, *Guide to the expression of uncertainty in measurement*. Int. Organ. Stand. Geneva (2024): 13.

ISO/IEC Guide 98-3:2008, *Uncertainty of measurement Part 3: Guide to the expression of uncertainty in measurement (GUM:1995)*. Int. Organ. Stand. Geneva (2008): 120.

JCGM 100:2008, *Evaluation of measurement data—Guide to the expression of uncertainty in measurement*. Int. Organ. Stand. Geneva ISBN 50 (2008): 134.

JCGM 101:2008, *Evaluation of measurement data – Supplement 1 to the "Guide to the expression of uncertainty in measurement" – Propagation of distributions using a Monte Carlo method*. Int. Organ. Stand. Geneva (2008): 82.

JCGM 102:2011, *Evaluation of measurement data – Supplement 2 to the "Guide to the expression of uncertainty in measurement" – Extension to any number of output quantities*. Int. Organ. Stand. Geneva (2011): 72.

Taylor, B. N. and C. E. Kuyatt, *Guidelines for evaluating and expressing the uncertainty of NIST measurement results*, NIST Technical Note 1297, Rev. ed., National Institute of Standards and Technology, Gaithersburg, MD, 2008. [Online]. Available: https://doi.org/10.6028/NIST.TN.1297

Vasilevskyi, O. M., V. Y. Kucheruk, and Y. T. Volodarskyi, *Uncertainty of measurement, control and test results*, OLDI-PLUS (2020): 352, ISBN: 978-966-289-374-8.

2 Varieties of Measurement Uncertainty

2.1 SYSTEMATIZATION OF VARIETIES OF MEASUREMENT UNCERTAINTIES

Based on the source of measurement uncertainty, they can be systematized into the following varieties: instrumental, methodological, and subjective.

2.1.1 Instrumental Varieties of Measurement Uncertainty

Instrumental varieties of measurement uncertainty are those uncertainties arising from imperfections in the manufacturing technology of measuring instruments. These uncertainties may result from manufacturing or calibration inaccuracies, component parameter drifts due to ageing, and other related effects. Each measuring instrument is associated with its instrumental uncertainty component, which must be considered in evaluating the combined standard uncertainty of the measurement result. It is not generally possible to eliminate a measuring instrument's response to variations in quantities other than the measurand. In addition to its sensitivity to the measurand, a measuring instrument typically exhibits (albeit to a lesser extent) sensitivity to changes in environmental or operating conditions. Therefore, instrumental uncertainty varieties are typically categorized as basic or additional.

The basic instrumental uncertainty component refers to the uncertainty of the measuring instrument under normal operating conditions, as defined by the manufacturer. The set of influence quantities and their permissible ranges that define normal conditions is specified in the instrument's technical documentation (e.g., datasheet or user manual).

When measurements are conducted under conditions that differ from those deemed normal for a given instrument type, the instrument's actual performance may degrade, leading to increased uncertainty. Such uncertainties are referred to as additional instrumental uncertainty components, and they arise under permissible but non-standard operating conditions, which are also specified in the technical documentation.

The instrumental uncertainty component of a measuring instrument of a particular type must not exceed a specified limit, referred to as the basic maximum permissible instrumental uncertainty for that type. The actual instrumental uncertainty of a specific instrument unit is a random quantity and may vary—possibly even being negligible—but it must remain within this predefined limit. If this condition is not met, the instrument must be excluded from the measurement process.

DOI: 10.1201/9781003742708-2

Instrumental uncertainty varieties are typically classified into four main categories[1]:

1. Basic instrumental uncertainty arises from the deviation of the actual transfer function of the measuring instrument from its nominal transfer function under normal operating conditions.
2. Additional instrumental uncertainty arises from the measuring instrument's response to variations in external influence quantities and non-informative parameters of the input signal relative to their normal values.
3. Dynamic uncertainty arises from the measuring instrument's response to the frequency content of the input signal. It depends on the dynamic characteristics of the instrument and the spectral content of the measurand.
4. Interaction uncertainty arises from the interaction between the measuring instrument and the object under measurement, where the act of connecting the measuring instrument alters the measurand itself.

The accuracy class of a measuring instrument is also an instrumental uncertainty component. For example, if a measuring instrument has an accuracy class of 0.5 and a measurement range of 0–150 V, this implies that the indication error does not exceed $\gamma = 0.5\%$ of the full-scale value. In this case, the basic instrumental uncertainty component can be estimated using the equation:

$$u_B = \frac{\gamma[\%] V_R}{100\%\sqrt{3}}, \tag{2.1}$$

where V_R is the nominal (rated) voltage value (150 V).

By substituting the specified numerical values into Equation (2.1), the basic instrumental uncertainty component of the measuring instrument is calculated to be 0.43 V.

2.1.2 Methodological Varieties of Measurement Uncertainty

A methodological variety of measurement uncertainty refers to the uncertainty component arising from imperfections in the measurement method (the measurement equation) or from inaccuracies introduced using simplified equations. For example, in the indirect measurement of the cross-sectional area of a circular rod $S = \pi d^2/4$ by directly measuring its diameter and then calculating the area, the result will include a methodological uncertainty due to the limited number of significant digits and the approximated value of π used in calculations.

The key feature of methodological uncertainty components is their individuality—they depend not only on the specific measurement method or physical principle used but also on the particular implementation of that method. This component is typically not stated in regulatory or normative documentation. However, such documentation may include operational guidelines for a specific measuring instrument, which can help reduce or eliminate some methodological uncertainty sources.

Typical causes of methodological uncertainty components include[2,3,4]:

1. Influence of non-informative parameters of the measurement object that are not accounted for in the object's model.
2. Influence of the method used to apply the measuring instrument.
3. Influence of algorithms (equations) used to compute the measurement result.
4. Influence of the correctness of the calibration model used for the measuring instrument with working standards or certified reference materials.
5. Mismatch between the assumed model of the measurement object and the actual model describing the property being measured (model uncertainty).

Let us consider an example illustrating the presence of a methodological uncertainty component.

The density ρ of a porous body of complex shape is determined by an indirect method in two steps: first, the mass m of the body is measured using a balance, then its volume is determined by submerging the body in a liquid and measuring the displaced volume. Because the surface of the body is rough (porous), and the liquid used in the experiment fills the surface pores, the measured volume of the body will be slightly lower than the actual volume by an amount approximately equal to V_{pores}, that is, the measured value is $\left(V - V_{\text{pores}}\right)$. In this case, the methodological uncertainty component is evaluated using the equation:

$$u(\Delta\rho_m) = \frac{\rho\left(\frac{V_{\text{pores}}}{V - V_{\text{pores}}} - \frac{V_{\text{pores}}}{V}\right)}{\sqrt{12}}. \quad (2.2)$$

To reduce this methodological uncertainty component, it is advisable to use a liquid that does not penetrate or wet the material of the body.

2.1.3 Subjective Varieties of Measurement Uncertainty

Subjective varieties of measurement uncertainty (also referred to as observer or operator uncertainties) arise due to the characteristics of the individual performing the observation.

These uncertainty components are caused by factors such as:

1. The inertial properties of human sensory organs, for example, delays in recording the maximum deflection of an indicator in ballistic-type instruments.
2. Observer positioning and reference system limitations, including interpolation errors when a reading falls between two marked divisions.
3. The limited sensitivity range and nonlinearity of human perception.

For instance, when reading analog scale divisions, different operators may interpret the same pointer position differently. One may consistently underestimate the reading, while another may tend to overestimate it. This results in the appearance of a subjective uncertainty component.

This type of uncertainty is often associated with operator inattention or errors, such as:

- failing to zero the instrument before use,
- incorrectly identifying the scale division value,
- inaccurately estimating the resolution, and
- making mistakes when connecting the measuring instrument.

The magnitude of the subjective uncertainty component depends significantly on the operator's skill and experience.

Subjective measurement uncertainty is characteristic of analog measuring instruments only and can generally be categorized into two types:

1. Reading uncertainty that occurs due to the operator rounding off the value when reading it from an analog scale. For example, if a pointer lies at approximately 54.3 scale divisions, one operator might read it as 54.3, another as 54.0, a third as 54.5, and so on.
2. Parallax uncertainty that arises from the relative positioning of the operator's eye, the pointer, and the scale of the analog instrument. Such uncertainty depends on the viewing angle and is not fixed or predictable in advance. As such, it cannot be specified in the instrument's technical documentation.

In contrast, digital measuring instruments perform automatic rounding, and the resulting uncertainty is referred to as quantization uncertainty. This type of uncertainty is standardized and is typically specified in the instrument's technical documentation. The uncertainty introduced by rounding the measurement result can be approximately calculated using the following equation:

$$u_{B\text{rounding}} = \frac{\Delta n_{\text{rounding}}}{\sqrt{3}}, \tag{2.3}$$

where $\Delta n_{\text{rounding}}$ represents the significant digits that are discarded during rounding of the measurement result.

Reduction or elimination of the subjective component of measurement uncertainty can be achieved using special types of scales, such as mirror scales, by employing digital readouts, and through automation of the measurement result acquisition.

Thus, subjective components of measurement uncertainty currently cannot be quantitatively evaluated and therefore are not included in the mathematical model of combined measurement uncertainty. Nevertheless, they should be minimized or eliminated wherever possible, and the operator must always be aware of their potential influence when reading values from the scale of an analog measuring instrument.

2.2 HEISENBERG'S UNCERTAINTY PRINCIPLE

Heisenberg's uncertainty principle is a fundamental concept of quantum mechanics, stating that it is inherently impossible to simultaneously measure both the position and momentum of a quantum object with arbitrary precision. This statement applies not only to the act of measurement but also to the theoretical formulation of the quantum state of a system. It is impossible to construct a quantum state in which a system is simultaneously characterized by exact values of both position and momentum.

The uncertainty principle was formulated in 1927 by the German physicist Werner Heisenberg and became a key milestone in understanding atomic phenomena and in the development of quantum mechanics.[5]

As the measurement accuracy or the sensitivity of the measuring instrument increases, fundamental natural factors emerge that limit the reliability of the measurement. The existence of such factors leads to the natural (inherent) uncertainty of measurement, which defines the potential accuracy limit of the process.

This limitation arises from the discreteness of measured quantities (e.g., electric charge cannot be measured with greater precision than the elementary charge) and fluctuations due to the discrete nature of matter and energy. At this level, measurement accuracy is limited by the laws of quantum mechanics. A formal indication that the measurement process has reached the quantum-mechanical limit of accuracy is the appearance of terms in the mathematical description that cannot be neglected, such as Planck's constant $\left(h = 6.63 \cdot 10^{-34}\ \mathrm{J \cdot s}\right)$. One such factor is Heisenberg's uncertainty principle, which imposes a limit on the maximum possible precision with which the dynamical variables of a microscopic system can be simultaneously defined. According to this principle, an individual quantity can, in principle, be measured with arbitrary accuracy, but two quantities whose quantum-mechanical operators do not commute cannot be simultaneously measured with unlimited precision.

For example, the uncertainty relation between the position x and the corresponding momentum component p_x takes the form:

$$\Delta x \Delta p_x \geq \frac{\hbar}{2}, \tag{2.4}$$

where $\hbar = h/_{2\pi}$.

A similar uncertainty relation is formulated for another pair of conjugate quantities—energy E and time t:

$$\Delta E \Delta t \geq \frac{\hbar}{2}. \tag{2.5}$$

This relation (2.5) links the uncertainty in the energy ΔE of an atomic system with the uncertainty in the time Δt during which the energy is measured.

Since Planck's constant is extremely small, these expressions have no practical significance in macroscopic measurements. The uncertainties in the respective quantities lie far beyond the limits of experimental accuracy currently achievable.

In macroscopic measurements, the maximum achievable accuracy is also limited by statistical fluctuations around the mean value. If such fluctuations cannot be reduced under fixed external conditions, they are generally referred to as noise. The sources of noise can typically be divided into three main categories:

- Thermal fluctuations at non-zero temperatures.
- The corpuscular (particle) nature of matter and electricity.
- The uncertainty relations of quantum mechanics.

2.3 THE PRINCIPLE OF COMPLEMENTARITY AND THE UNCERTAINTY RELATIONS

The principle of complementarity emerged as an attempt to understand the origin of the seemingly contradictory visual representations associated with objects of the microworld.

In many experiments, electrons and other elementary particles exhibit corpuscular (particle-like) properties, meaning they behave as discrete, localized entities. Any device used to detect micro-objects always registers them as whole objects, localized within a very small region of space.

On the other hand, during motion, all microparticles exhibit wave-like properties. Interference (the superposition of waves) and diffraction (the bending of waves around obstacles) are observed when particles interact with crystalline lattices or artificially created barriers. Electrons and other particles behave like waves that bend around obstacles and seemingly pass through multiple slits in diffraction gratings simultaneously.

Thus, all micro-objects exhibit wave–particle duality. A general answer to how these contradictory properties coexist in a single object was provided by Niels Bohr. Bohr emphasized that all instruments used to record individual events in the microworld are inherently macroscopic and cannot be otherwise. Our sensory organs cannot directly perceive microprocesses; humans themselves are macroscopic beings. Therefore, the concepts we use to describe physical phenomena are macroscopic in nature and reflect how instruments operate. However, these concepts cannot be fully applied to micro-objects, as their behavior does not follow the laws of classical mechanics. According to Bohr's principle of complementarity, a complete description of quantum-mechanical phenomena requires the use of two mutually exclusive (complementary) sets of classical concepts—particle and wave. Only the totality of these concepts provides a comprehensive understanding of a physical phenomenon.

The principle of complementarity is the result of a philosophical interpretation of the fundamentally new physical theory—quantum mechanics. It expresses, on the macroscopic level, one of the core dialectical laws—the unity of opposites. A particular manifestation of the principle of complementarity is Heisenberg's uncertainty principle.

When we speak of particles, we imagine matter located at a definite point in space, having a certain energy and moving with a well-defined velocity. We implicitly assume that it is possible to know a particle's position, momentum, and energy with complete precision at any given moment. However, when associating a particle with a wave, we switch to the image of an infinite sinusoidal function spread out in space. In such a context, notions like "wavelength at a specific point" or "momentum at a specific point" lose their meaning. Similarly, the concept of the particle's energy at a specific moment in time becomes meaningless. According to Planck's formula, energy is linked to the frequency of a wave that characterizes a harmonic oscillation process in time. The statement that an electron can only approximately be considered a material point means that its position, momentum, and energy can only be defined approximately. Quantitatively, this is expressed by Heisenberg's uncertainty relation (Equation 2.5). According to this relation, the more precisely the momentum of a particle is known, the greater the uncertainty in its position. A similar relation holds between energy and time: the precision of energy measurement is inversely proportional to the duration of the measurement process. This is due to the interaction between the measuring instrument and the measured object.

The uncertainty principle also explains why an electron cannot "fall" into the nucleus of an atom. The atomic nucleus is extremely small, and if the electron were to fall into it, its location would be determined with very high precision. As a result, the uncertainty in the electron's momentum would increase dramatically, leading to a large spread in possible velocities. Among these, some velocities would be so high that the electron would be more likely to escape the atom than fall into the nucleus.

2.4 THE PRINCIPLE OF SUPERPOSITION

The principle of superposition is one of the most general and fundamental laws in many areas of physics, especially in quantum mechanics. This principle states that the resulting effect of several influences acting simultaneously is equal to the sum of the individual effects produced by each influence acting alone. One simple example is the parallelogram rule used to determine the net force acting on a body due to two individual forces. The principle of superposition holds only when the acting phenomena do not influence each other. For instance, a headwind slows down a vehicle in accordance with the parallelogram rule—in this case, the principle of superposition applies fully. However, if wind-blown sand damages the engine and affects its performance, the superposition principle no longer holds, as the phenomena are no longer independent.

In classical Newtonian physics, the superposition principle is not universal; it often holds only as an approximation in many situations.

In contrast, in the microscopic (quantum) world, the principle of superposition is a foundational concept. Alongside the uncertainty principle, it forms the mathematical basis of quantum mechanics. Unlike classical physics, the superposition principle in quantum theory lacks intuitive visual representation, because it involves the addition of mutually exclusive (alternative) quantum states.

In relativistic quantum theory, where particle transformations are allowed (such as creation and annihilation), the superposition principle must be supplemented by the superselection rule. For example, in electron–positron annihilation, the principle of superposition is complemented by the conservation of electric charge—the total charge before and after the interaction must remain the same. Since the charges of the electron and positron are equal and opposite, the result of their annihilation must be a neutral particle, which in this case is a photon.

Perhaps the most well-known manifestation of the superposition principle is in electrostatics, where it states that the electric field intensity at a given point in space due to a system of charges is the vector sum of the field intensities due to each individual charge.

The principle of superposition can also be expressed through several equivalent formulations:

- The interaction between two particles remains unchanged upon introducing a third particle that also interacts with the first two.
- The total interaction energy in a multi-particle system is the sum of pairwise interaction energies between all possible particle pairs—there are no true many-body interactions.
- The equations describing multi-particle systems are linear with respect to the number of particles.

It is the linearity of a fundamental theory in a given domain of science or engineering that gives rise to the superposition principle.

In the quantum realm, the superposition principle is a cornerstone of quantum mechanics, together with Heisenberg's uncertainty principle, and it defines the structure and behavior of quantum systems.

2.5 CRITERION FOR NEGLIGIBILITY OF MEASUREMENT UNCERTAINTY

Calculations can be simplified when expressing the combined measurement uncertainty by neglecting uncertainty components with minimal (very small) values. To do this, a criterion for negligible uncertainty must be established—that is, a mathematical rule must be defined which allows one to disregard those uncertainty contributions that have an insignificant effect.

Let us consider the criterion for negligible uncertainty using the example of indirect measurements. In the simplest case, the combined standard uncertainty is determined using the following equation:

$$u_c = \sqrt{u_1^2 + u_2^2}, \tag{2.6}$$

where u_1 is the standard uncertainty with the greater value, and u_2 is the standard uncertainty with the lesser value.[6]

When rounding an uncertainty, its second significant digit is typically discarded. A smaller standard uncertainty u_2 may be neglected if its inclusion does not affect the first significant digit of the combined uncertainty. For this, the following condition must be satisfied:

$$u_c - u_1 \leq 0.05 u_c. \tag{2.7}$$

From Equation (2.6), we obtain:

$$u_1^2 + u_2^2 = u_c^2; \tag{2.8}$$

$$u_c^2 - u_1^2 = u_2^2. \tag{2.9}$$

Hence, by dividing expression (2.9) by expression (2.7), we obtain:

$$u_c + u_1 \geq \frac{u_2^2}{0.05u_c}. \tag{2.10}$$

Assuming that $u_c \approx u_1$, we obtain:

$$u_2 \leq \sqrt{2 \cdot 0.05u_1^2} = 0.3u_1. \tag{2.11}$$

Thus, if the smaller standard uncertainty u_2 is at least three times smaller than the larger standard uncertainty u_1, it can be neglected. Equation (2.11) represents the criterion for negligible uncertainty. A group of uncertainties can be disregarded if their sum is less than one third of the largest uncertainty.

Notes

1 O. M. Vasilevskyi, V. Yu. Kucheruk, Ye. T. Volodarskyi, *Uncertainty of measurement, control and test results* (OLDI-PLUS, 2020: 352).
2 ISO/IEC Guide 98-1:2024, *Guide to the expression of uncertainty in measurement* (Int. Organ. Stand., Geneva, 2024: 13).
3 ISO/IEC Guide 98-3:2008, *Uncertainty of measurement Part 3: Guide to the expression of uncertainty in measurement* (GUM:1995) (Int. Organ. Stand., Geneva, 2008: 120).
4 JCGM 100:2008, *Evaluation of measurement data—Guide to the expression of uncertainty in measurement* (Int. Organ. Stand., Geneva, ISBN 50, 2008: 134).
5 Messiah Albert, *Quantum mechanics*. Vol. I & II. (North-Holland Publishing Company, Amsterdam/Elsevier Science B.V., 1961).
6 Ian Farrance, Tony Badrick, Robert Frenkel, Uncertainty in measurement: A review of the procedures for determining uncertainty in measurement and its use in deriving the biological variation of the estimated glomerular filtration rate, *Practical Laboratory Medicine*, Volume 12, 2018, e00097.

BIBLIOGRAPHY

Albert, Messiah, *Quantum mechanics*. vol. I & II. North-Holland Publishing Company, Amsterdam/Elsevier Science B.V., 1961. ISBN 978-0720400458.

Farrance, Ian, Tony Badrick, and Robert Frenkel, Uncertainty in measurement: A review of the procedures for determining uncertainty in measurement and its use in deriving the biological variation of the estimated glomerular filtration rate, *Practical Laboratory Medicine*, 12, 2018, e00097, https://doi.org/10.1016/j.PLABM.2018.e00097

ISO/IEC Guide 98-1:2024, *Guide to the expression of uncertainty in measurement*, Int. Organ. Stand. Geneva (2024): 13.

ISO/IEC Guide 98-3:2008, *Uncertainty of measurement Part 3: Guide to the expression of uncertainty in measurement (GUM:1995)*, Int. Organ. Stand. Geneva (2008): 120.

JCGM 101:2008, *Evaluation of measurement data – Supplement 1 to the "Guide to the expression of uncertainty in measurement" – Propagation of distributions using a Monte Carlo method*, Int. Organ. Stand. Geneva (2008): 82.

Vasilevskyi, O. M., V. Y. Kucheruk, and Y. T. Volodarskyi, *Uncertainty of measurement, control and test results*, OLDI-PLUS (2020): 352, ISBN: 978-966-289-374-8.

3 Evaluation of Research Results Based on the Concept of Measurement Uncertainty

3.1 EVALUATION OF DIRECT MEASUREMENT RESULTS

Direct measurement is the measurement of a single physical quantity, the value of which is obtained directly without transformation of its nature or the use of known dependencies.

The measured quantity Y is assumed equal to the quantity X that is directly measured, $Y = X$. To perform direct measurements of the physical quantity X, it is necessary to have a comparator and a multivalued measure with an appropriate range of values, or a single-valued measure and a scale-type measuring transducer.

In practice, the most common are measurements with single observations, which are the simplest, most efficient, and least costly.

3.1.1 Evaluation of Direct Measurement Results with Single Observations

Direct single measurements are performed once, and the value of the physical quantity is determined based on the readings of the measuring instrument intended for measuring that quantity. However, in some cases, direct single measurements are repeated several times. Redundant measurements are carried out to protect against equipment malfunctions, abnormal results, etc., and the final result is determined by selecting one measurement from the series of conducted measurements.

The estimate of the measured quantity Y is directly given by the instrument reading. Therefore, data processing in this case involves analysis of the components of measurement uncertainty, elimination of systematic components of uncertainty, evaluation of residual systematic uncertainty components that are not eliminated, and summation of uncertainty components.

Thus, to evaluate the result of a direct measurement with a single observation, uncertainty evaluation methods of Type B are used (see Sections 1.3.2 and 1.4.1). All uncertainty components assessed as Type B are summed to determine the combined uncertainty (see Section 1.4.2 or 1.4.3). In cases where an interval estimate

DOI: 10.1201/9781003742708-3

of uncertainty is required, the expanded uncertainty is additionally determined (see Section 1.4.4).

For example, A voltage measurement was carried out using a voltmeter with an upper measurement limit of $V_{\max} = 1000$ V and an accuracy class of 0.1/0.01. As a result of the measurement, the voltmeter reading was $V_m = 500.2$ V. The measurement was performed at an ambient temperature of $T = 26^{\circ}$C and in the presence of a magnetic field with an intensity of $H = 400$ A/m.It is known that the temperature error does not exceed the basic error for every $\Delta T = 10^{\circ}$C deviation from the reference temperature of 20 °C, and the magnetic error amounts to half of the basic instrumental error of the measuring instrument when the external magnetic field intensity is up to 400 A/m.

It is required to estimate the uncertainty of the voltage measurement result. Measurement uncertainty consists of the basic instrumental uncertainty, which is determined according to the voltmeter's accuracy class, and additional uncertainty components caused by deviations of ambient temperature from standard conditions and the presence of an external magnetic field.

Since the instrument accuracy class is defined by the parameters *c* (the sum of the reduced additive and relative multiplicative errors) and *d* (the reduced additive error), namely 0.1/0.01, the equation for calculating the accuracy class has the following form:

$$\gamma = \pm\left[\mathrm{c} + d\left(\frac{V_{\max}}{V_m} - 1\right)\right] = \pm\left[0.1 + 0.01\left(\frac{1000}{500.2} - 1\right)\right] = \pm 0.11. \quad (3.1)$$

The basic instrumental uncertainty of the voltage measurement (Type B uncertainty) in this case is calculated similarly to Equation (2.1) and is equal to:

$$u_{MI} = \frac{U\gamma}{100\%\sqrt{3}} = 500.2\frac{0.11}{1.73 \cdot 100} \approx 0.32 \text{ V}. \quad (3.2)$$

Since the standard ambient temperature is 20 °C, and the ambient temperature at which the measurement was performed was $T = 26^{\circ}$C, the additional uncertainty due to temperature deviation is equal to:

$$u_T = \frac{T - T_{\text{standard}}}{\Delta T} u_{MI} = \frac{26 - 20}{10} 0.32 \approx 0.19 \text{ V}. \quad (3.3)$$

The additional uncertainty caused by the influence of the external magnetic field intensity $H = 400$ A/m is equal to:

$$u_H = \frac{1}{2} u_{MI} = 0.16 \text{ V}. \quad (3.4)$$

The combined uncertainty of the voltage measurement is equal to:

$$u_c = \sqrt{u_{MI}^2 + u_T^2 + u_H^2} = 0.41 \text{ V}. \quad (3.5)$$

Assuming a level of confidence $p = 0.95$ (based on the assumption of a normal distribution of the measurement result, with a coverage factor $k = 1.96$), the expanded uncertainty is calculated as:

$$U = ku_c = 1.96 \cdot 0.41 = 0.8 \text{ V}. \tag{3.6}$$

Thus, the measurement result is expressed as:

$$V_m = 500.2 \pm 0.8 \text{ V}, \text{at } p = 0.95 \text{ or } 499.4 \leq V_m \leq 501 \text{ V}, \text{at } p = 0.95. \tag{3.7}$$

Thus, when evaluating the results of direct measurements with single observations, various cases may occur depending on the number of uncertainty components. There may be only one basic uncertainty, or there may be several uncertainty components. Uncorrected components of standard uncertainties that remain after applying corrections are summed.

3.1.2 Evaluation of Direct Measurement Results with Multiple Observations

Direct multiple measurements are measurements in which the estimate of the measured quantity is obtained based on a series of results from direct measurements.

In the general case, when processing the results of direct measurements with multiple observations, the following steps should be performed[1]:

1. Eliminate systematic effects that lead to standard uncertainty components.
2. Calculate the estimate of the measured quantity:

$$y = \bar{y} = \sum\nolimits_{k=1}^{N} \frac{y_k}{N}, \tag{3.8}$$

   where y_k are the results of individual observations ($k = 1, 2, \ldots, N$).
3. Estimate the Type A uncertainty of the measurement result:

$$u_A = \sqrt{\frac{\sum_{k=1}^{N} (y_k - \bar{y})^2}{N - 1}}. \tag{3.9}$$

4. Eliminate outliers from the measurement results. To do this, the following ratio is calculated[2]:

$$s_1 = \frac{\bar{y} - y_{\min}}{u_A}; \tag{3.10}$$

$$s_2 = \frac{y_{\max} - \bar{y}}{u_A}. \tag{3.11}$$

   Then, by specifying the confidence probability p, the reliability is calculated using the following equation: $\alpha = 1 - p$. Depending on the

TABLE 3.1
Acceptable Values of Normalized Deviations

Sample Size, N	Reliability, α			
	0.100	0.075	0.050	0.025
3	1.15	1.15	1.15	1.15
4	1.42	1.44	1.46	1.48
5	1.60	1.64	1.67	1.72
6	1.73	1.77	1.82	1.89
7	1.83	1.88	1.94	2.02
8	1.91	1.96	2.03	2.13
9	1.98	2.04	2.11	2.21
10	2.03	2.10	2.18	2.29
11	2.09	2.14	2.23	2.36
12	2.13	2.20	2.29	2.41
13	2.17	2.24	2.33	2.47
14	2.21	2.28	2.37	2.50
15	2.25	2.32	2.41	2.55
16	2.28	2.35	2.44	2.58
17	2.31	2.38	2.48	2.62
18	2.34	2.41	2.50	2.66
19	2.36	2.44	2.53	2.68
20	2.38	2.46	2.56	2.71

number of measurements N, the permissible value s_{crit} is selected from Table 3.1. If the calculated values s_1 and s_2 are both less than the permissible value s_{crit}, the hypothesis of the presence of outliers in the observation series is rejected. However, if either s_1 or s_2 exceeds s_{crit}, this indicates the presence of outliers in the measurement results. These outliers must be eliminated from the data set.

5. Estimate the Type A uncertainty of the arithmetic mean:

$$u_A(\bar{y}) = \sqrt{\frac{\sum_{k=1}^{N}(y_k - \bar{y})^2}{N(N-1)}}. \tag{3.12}$$

6. Estimate the components $u_i(Y)$ of the combined uncertainty of Type B associated with uncorrected residual standard uncertainties (see Section 1.4.1).
7. Calculate the combined Type B uncertainty:

$$u_{cB}(Y) = \sqrt{\sum_{i=1}^{m} c_i^2 u_i^2(Y)}. \tag{3.13}$$

8. Estimate the combined uncertainty of the measurement result:

$$u_c(Y) = \sqrt{u_A^2(\bar{y}) + u_{cB}^2(Y)}. \tag{3.14}$$

9. Estimate the expanded uncertainty of the measurement result:

$$U = t_p\left(v_{eff}\right)u_c(Y). \tag{3.15}$$

10. Report the measurement result with the specified expanded uncertainty:

$$Y = \bar{y} \pm U, \text{for } p, \tag{3.16}$$

or

$$\bar{y} - U \leq Y \leq \bar{y} + U, \text{for } p. \tag{3.17}$$

If the measurement cannot be repeated, in addition to the expanded uncertainty at the level of confidence of $p = 0.95$, the expanded uncertainty may also be reported for the level of confidence of $p = 0.99$.

For measurements performed in medicine and environmental science (critical for human health), it is permissible to select a higher confidence level (e.g., 0.9973 or 0.999) instead of $p = 0.99$.

3.1.3 Evaluation of Groups of Direct Measurements with Multiple Observations

If multiple measurements are conducted over a long period, it is necessary to consider that changes in the measuring instrument parameters and the surrounding environment may cause systematic or random variations in the estimate of the measurement result and its standard deviation (SD). To minimize these variations, measurements are performed in several series, with the measuring equipment being recalibrated in each series and measures taken to stabilize environmental parameters.

When several groups of direct multiple measurements of the same physical quantity are available, it is important to ensure that additional systematic uncertainties are absent and that the SD estimates are close.

If there are no systematic uncertainty components in the measurement groups (i.e., the difference between the arithmetic mean values is acceptable) and the differences between SD estimates are acceptable, such groups are called groups with homogeneous variances. This means that the arithmetic mean values and variance estimates in all groups are estimates of a single true value and a single variance.

Assume there are G groups of direct multiple measurements of quantity Y with N observations in each group. The methodology for processing groups of direct measurements with multiple observations is as follows.

The arithmetic mean values of each observation group are determined by the equation:

$$\bar{y}_G = \frac{1}{N}\sum_{n=1}^{N} y_{Gn}, \tag{3.18}$$

where n is the number of observations of the quantity $Y (n = 1, 2, \ldots, N)$ in the Gth group $(G = 1, 2, \ldots, g)$.

The total arithmetic mean $\bar{y}$ of the group means $\bar{y}_G$, accepted as the best estimate of the measured quantity Y, is determined by the equation:

$$Y = \bar{y} = \frac{1}{G}\sum_{G=1}^{g} \bar{y}_G = \frac{1}{GN}\sum_{G=1}^{g}\sum_{n=1}^{N} y_{Gn}. \tag{3.19}$$

The within-group variance in the Gth group is determined by the equation[3]:

$$s_G^2 = \frac{1}{N-1}\sum_{n=1}^{N}(y_{Gn} - \bar{y}_G)^2. \tag{3.20}$$

The total variability of the measurement results, that is, the total variance of the arithmetic means, is determined by the equation:

$$s^2(\bar{y}_G) = \frac{1}{G-1}\sum_{G=1}^{g}(\bar{y}_G - \bar{y})^2. \tag{3.21}$$

Next, it is determined whether the between-group component of the variance is significant compared to the within-group component.

To do this, the following operations are performed:

1. Two independent estimates of the averaged within-group variance of observations are determined. The first estimate, denoted as s_I^2, is obtained from the observed deviations of the daily arithmetic means $\bar{y}_G$ (Equation 3.18). Since $\bar{y}_G$ is the arithmetic mean of N observations, its estimated total variance $s^2(\bar{y}_G)$, under the assumption that the between-group variance is zero, is estimated as $\frac{s_I^2}{N}$. Then, from Equation (3.20), it follows that:

$$s_I^2 = N s^2(\bar{y}_G) = \frac{N}{G-1}\sum_{G=1}^{g}(\bar{y}_G - \bar{y})^2. \tag{3.22}$$

The variance estimate (3.22) has $(G - 1)$ degrees of freedom.

The second estimate, denoted as s_{II}^2, is the average of the variances obtained from the G individual within-group variance values $s_G^2(y_{Gn})$:

$$s_{II}^2 = \bar{s}_G^2(y_{Gn}) = \frac{\sum_{G=1}^{g} \nu_G s_G^2(y_{Gn})}{\sum_{G=1}^{g} \nu_G}. \tag{3.23}$$

Since the structure is balanced and all degrees of freedom are equal $\nu_G = N - 1$, the resulting expression for s_{II}^2 is simply the arithmetic mean of $s_G^2(y_{Gn})$:

$$s_{II}^2 = \bar{s}_G^2(y_{Gn}) = \frac{1}{G}\sum\nolimits_{G=1}^{g} s_G^2(y_{Gn}). \tag{3.24}$$

Thus, considering expression (3.20), we obtain:

$$s_{II}^2 = \frac{1}{G(N-1)}\sum\nolimits_{G=1}^{g}\sum\nolimits_{n=1}^{N}(y_{Gn} - \bar{y})^2, \tag{3.25}$$

which is an estimate with $G(N-1)$ degrees of freedom.

Since the estimate s_I^2 is based on the variability of the arithmetic means, whereas the estimate s_{II}^2 is based on the variability of within-group observations, their difference indicates the possible presence of between-group variability.

2. The values of s_I^2 and s_{II}^2 are compared. For this purpose, the ***F*-test** is used. It is known that the F-distribution of components is the probability distribution of the ratio

$$F(\nu_I, \nu_{II}) = \frac{s_I^2(\nu_I)}{s_{II}^2(\nu_{II})} \tag{3.26}$$

of two independent estimates $s_I^2(\nu_I)$ and $s_{II}^2(\nu_{II})$ of the variance of a normally distributed random variable. The parameters ν_I and ν_{II} are the corresponding degrees of freedom of the two estimates, and $0 \leq F(\nu_I, \nu_{II}) \leq \infty$. The critical values of F for different probabilities (i.e., quantiles of the F-distribution) are given in the Fisher distribution table for various values of ν_I and ν_{II}.[4]

   Typically, critical values of F are given for probabilities of 0.95, 0.975, or 0.99.

If the calculated value $F(\nu_I, \nu_{II})$ is greater than the critical value for the selected probability level ($F_{0.95}, F_{0.975}$ or $F_{0.99}$), this is interpreted as evidence of the presence of between-group variance, since $s_I^2(\nu_I)$ exceeds $s_{II}^2(\nu_{II})$ by a statistically significant amount.

If $F(\nu_I, \nu_{II}) < F_p$, the existence of between-group uncertainty is rejected, as the difference between s_I^2 and s_{II}^2 is not considered statistically significant. In this case, the estimated variance $s^2(\bar{y})$ for $\bar{y}$ should be calculated using the general expression:

$$s^2(\bar{y}) = \frac{1}{GN(GN-1)}\sum\nolimits_{G=1}^{g}\sum\nolimits_{n=1}^{N}(y_{Gn} - \bar{y}_G)^2. \tag{3.27}$$

Expression (3.27) is equivalent to the following equation:

$$s^2(\bar{y}) = \frac{(G-1)s_I^2 + G(N-1)s_{II}^2}{GN(GN-1)}. \tag{3.28}$$

Expressions (3.27) and (3.28) can be represented as:

$$s^2(\bar{y}) = \frac{1}{GN(GN-1)}\sum_{G=1}^{g}\sum_{n=1}^{N}[(y_{Gn} - \bar{y}_G) + (\bar{y}_G - \bar{y})]^2 =$$

$$= \frac{1}{GN(GN-1)}\sum_{G=1}^{g}\sum_{n=1}^{N}\left[(y_{Gn} - \bar{y}_G)^2 + (\bar{y}_G - \bar{y})^2 + 2(y_{Gn} - \bar{y}_G)(\bar{y}_G - \bar{y})\right], \tag{3.29}$$

where $\sum_{G=1}^{g}\sum_{n=1}^{N}(y_{Gn} - \bar{y}_G)^2 = G(N-1)s_{II}^2$; $\sum_{G=1}^{g}\sum_{n=1}^{N}(y_G - \bar{y})^2 = G(N-1)s_I^2$; $\sum_{G=1}^{g}\sum_{n=1}^{N}2(y_{Gn} - \bar{y}_G)(\bar{y}_G - \bar{y}) = 2\sum_{G=1}^{g}(\bar{y}_G - \bar{y})\sum_{n=1}^{N}(y_{Gn} - \bar{y}_G) = 2\sum_{G=1}^{g}(\bar{y}_G - \bar{y})(N\bar{y}_G - N\bar{y}_G) = 0$.

Thus, assuming that all corrections for systematic effects have been considered and all other components of uncertainty are negligible, the measurement result can be written as:

$$Y = \frac{1}{GN}\sum_{G=1}^{g}\sum_{n=1}^{N}y_{Gn} = \frac{1}{G}\sum_{G=1}^{g}\sum_{n=1}^{N}\bar{y}_G = \bar{y}, \tag{3.30}$$

with the combined standard uncertainty (3.28), which has $GN - 1$ degrees of freedom.

The expanded uncertainty of groups of repeated observations is calculated using the equation:

$$U = t_p(\nu)s(\bar{y}), \tag{3.31}$$

where $t_p(\nu)$ is Student's t-factor for the number of degrees of freedom $\nu = GN - 1$ and confidence level p. For $\nu > 30$, the coverage factor for the normal distribution can be used instead of Student's t-factor.

If $F(\nu_I, \nu_{II}) \geq F_p$, then the existence of inter-group variance is accepted and assumed to be random. In that case, the estimated variance $\bar{y}$ is derived from $s^2(\bar{y}_G)$, as it properly reflects both intra-group and inter-group random components of variance. It is calculated using the equation:

$$s^2(\bar{y}) = \frac{s^2(\bar{y}_G)}{G}. \tag{3.32}$$

Considering Equation (3.21), we obtain:

$$s^2(\bar{y}) = \frac{\sum_{G=1}^{g}(\bar{y}_G - \bar{y})^2}{G(G-1)}. \tag{3.33}$$

The variance estimate described by expression (3.33) has $G-1$ degrees of freedom.

In this case, the expanded uncertainty of the measurement result is calculated using Equation (3.31), but with the number of degrees of freedom $v = G - 1$.

3.2 EVALUATION OF INDIRECT MEASUREMENT RESULTS

Indirect measurements are understood as indirect determinations of a quantity obtained by transforming its nature or by calculation based on the results of direct single measurements of other physical quantities, with which the measured quantity is explicitly related through a known functional dependence.

The dependence of the indirectly measured quantity Y on the arguments $X_1, X_2, \ldots, X_N$ is known

$$Y = f(X_1, X_2, \ldots, X_N). \tag{3.34}$$

The measurement results of the arguments must be consistent with each other, meaning that the values of the arguments obtained from direct measurements must correspond to the same value of the measured quantity. Failure to meet this condition is a source of methodological uncertainty.

The processing of results from indirect measurements is determined by:

- the form in which the measurement results of the arguments are presented;
- the type of the transformation equation dependence.

According to the type of functional dependence *f*, they are divided into two main groups—linear and nonlinear.

In the case of a nonlinear functional dependence *f* or the presence of correlation among the estimates $x_1, x_2, \ldots, x_N$, reliable estimation of the measured quantity should be performed using the equation:

$$y = \bar{Y} = \frac{1}{n}\sum_{k=1}^{n} f(X_{1,k}, X_{2,k}, \ldots, X_{N,k}), \tag{3.35}$$

where $X_{1,k}, X_{2,k}, \ldots, X_{N,k}$ is the *k*th observation $(k = 1, 2, \ldots, n)$ of the input quantities $X_1, X_2, \ldots, X_N$.

Thus, y is taken as the mean value of n independent observations Y_k of the quantity Y; each observation has the same uncertainty and is based on a complete set of observed values of the N input quantities X_i, obtained at the same time.

In this case, the Type A uncertainty of the result of the indirect measurement is calculated using the equation:

$$u_A = \sqrt{\frac{\sum_{k=1}^{n} (Y_k - y)^2}{n(n-1)}}. \tag{3.36}$$

In practice, the estimate y of the measured quantity Y is usually obtained from a simpler expression:

$$y = (\bar{X}_1, \bar{X}_2, \ldots, \bar{X}_N), \tag{3.37}$$

where $\bar{X}_i = \sum_{k=1}^{n} (p_{i,k} X_{i,k})$ is the arithmetic mean of individual observations $X_{i,k}$, obtained with probabilities $p_{i,k}$.

In this case, the estimated standard deviation associated with the output estimate or with the measurement result y is called the combined standard uncertainty and is denoted as $u_c(y)$. It consists of the estimated standard deviations associated with each input estimate x_i, which are called standard uncertainties and are denoted as $u(x_i)$.

Each input estimate x_i and its associated standard uncertainty $u(x_i)$ are obtained from the probability distribution of possible values of the input quantity X_i. This probability distribution, as previously mentioned, may be based either on a series of observations $X_{i,k}$ of the quantity X_i or on priori information. In the first case, the components of standard uncertainty are evaluated as Type A estimates, while in the second case—as Type B estimates.

The method of combining standard uncertainties depends on the degree of correlation between the input quantities.

3.2.1 Evaluation of Uncorrelated Input Quantities

In the absence of correlation between the input quantities, the procedure for processing the results of indirect measurements is as follows.

1. The estimate of the measured quantity Y is determined using the equation $y = f(x_1, x_2, \ldots, x_N)$, where $x_1, x_2, \ldots, x_N$ are the estimates of the input quantities $X_1, X_2, \ldots, X_N$, obtained from the results of single or repeated direct measurements. In the latter case, the estimates are taken as the arithmetic mean of n observations of X_i

$$x_i = \bar{X}_i = \sum_{n=1}^{N} (p_{i,n} x_{i,n}). \tag{3.38}$$

2. The standard uncertainties $u(x_i)$ of the estimates of the input quantities $x_1, x_2, \ldots, x_N$ are determined. These may be evaluated by Type A methods (only in the case of repeated measurements of $X_1, X_2, \ldots, X_n$) or by Type B methods (in the case of both single and repeated measurements of $X_1, X_2, \ldots, X_n$), as described earlier.
3. The combined standard uncertainty of the estimate of the measurand y is calculated using the following equation:

$$u_c(y) = \sqrt{\sum_{i=1}^{N} u_i^2(y)} = \sqrt{\sum_{i=1}^{N} \left(\frac{\partial f}{\partial x_i}\right)^2 u^2(x_i)}, \tag{3.39}$$

where $\frac{\partial f}{\partial x_i} = c_i$ are the sensitivity coefficients, calculated as the partial derivatives $\frac{\partial f}{\partial X_i}$ evaluated at $X_i = x_i$.

If the function f is significantly nonlinear, higher-order terms of the Taylor series expansion should be included in expression (3.39). If the probability distribution of each X_i is symmetric about its mean value, it is sufficient to include only the second-order terms under the square root:

$$u_c^2(y) = \sum_{i=1}^{N} \left(\frac{\partial f}{\partial x_i}\right)^2 u^2(x_i) + $$
$$+ \sum_{i=1}^{N} \sum_{j=1}^{N} \left[\frac{1}{2}\left(\frac{\partial^2 f}{\partial x_i \partial x_j}\right)^2 + \frac{\partial f}{\partial x_i}\frac{\partial^3 f}{\partial x_i \partial x_j^2}\right] u^2(x_i) u^2(x_j). \tag{3.40}$$

4. The expanded uncertainty of the estimate of the measurand is determined using the following equation:

$$U = k \cdot u_c(y), \tag{3.41}$$

where the coverage factor k is taken as Student's t-factor $t_p(v_{\text{eff}})$, corresponding to the effective degrees of freedom v_{eff}, obtained from the Welch–Satterthwaite equation:

$$v_{\text{eff}} = \frac{u_c^4(y)}{\sum_{i=1}^{N} \frac{u_i^4(y)}{v_i}} = \frac{\left[\sum_{i=1}^{N} u_i^2(y)\right]^2}{\sum_{i=1}^{N} \frac{u_i^4(y)}{v_i}}. \tag{3.42}$$

In Equation (3.42) $u_i(y) = \frac{\partial f}{\partial x_i} u(x_i)$, v_i is taken as ∞ if $u(x_i)$ is evaluated by a Type B method, and as $n - 1$ if $u(x_i)$ is evaluated by a Type A method, where n is the number of repeated observations used to obtain the estimate of x_i.

5. The measurement result is expressed in the form:

$$Y = y \pm U, \quad p. \tag{3.43}$$

3.2.2 Evaluation of Correlated Input Quantities

When there is a correlation between input quantities, the procedure for processing the results of indirect measurements is as follows.

1. The estimate of the measurand is determined using an equation $y = f(x_1, x_2, \ldots, x_N)$ involving the estimates of the input quantities $X_1, X_2, \ldots, X_N$, which are obtained from repeated direct measurements. These estimates are taken as the arithmetic means $\bar{X}_i = \sum_{k=1}^{N} p_{i,k} x_{i,k}$ of N observations. For measurements of equal precision, the mean value is calculated using the equation $\bar{X}_i = \frac{1}{n}\sum_{k=1}^{N} x_{i,k}$. As in the case of independent indirect measurements, it is assumed that any observations affected by gross errors or outliers have been removed from the set of repeated observations, and that all known corrections for systematic effects have been applied.

2. The standard uncertainties $u(x_i)$ of the input quantity estimates are determined. These uncertainties can be obtained as Type A evaluations (only in the case of repeated measurements of $X_1, X_2, \ldots, X_N$) or as Type B evaluations (applicable in both single and repeated measurements of $X_1, X_2, \ldots, X_N$).
3. Calculate the sensitivity coefficients

$$c_i = \frac{\partial f}{\partial X_i} \text{ for } X_i = x_i \ (i = 1, 2, \ldots, N). \tag{3.44}$$

4. Determine the pairwise estimates of the covariance terms

$$u(x_i, x_j) = \frac{\sum_{k=1}^{n} (x_{i,k} - \bar{x}_i)(x_{j,k} - \bar{x}_j)}{n - 1}. \tag{3.45}$$

5. Calculate the correlation coefficient

$$r(x_i, x_j) = \frac{u(x_i, x_j)}{u(x_i) u(x_j)}, \quad -1 < r(x_i, x_j) < 1. \tag{3.46}$$

6. Determine the estimate of the variance of the measurement result

$$u_c^2 = \sum_{i=1}^{N} c_i^2 u^2(\bar{x}_i) + 2 \sum_{j=1}^{N-1} \sum_{i=2}^{N} c_i c_j r(x_i, x_j) u(x_i) u(x_j). \tag{3.47}$$

7. Determine the expanded uncertainty of the indirect measurement result $U = t_p(\nu_{\text{eff}}) u_c$, where $t_p(\nu_{\text{eff}})$ is Student's t-factor for the given confidence level p and degrees of freedom ν_{eff}, which is determined according to Equation (3.42).
8. Record the measurement results in the form of

$$Y = y \pm U, \quad p \text{ or } y - U \leq Y \leq y + U, \quad p.$$

In practice, situations may arise where information is lacking to accurately estimate correlation coefficients or covariances. If such information cannot be obtained within acceptable resource and time constraints, approximate estimates must be used. For example, if no information is available about the values or signs of the correlation coefficients, and both overestimation and underestimation of uncertainty are equally undesirable, their values are assumed to be zero. However, if underestimation of uncertainty is unacceptable, instead of quadratic summation, linear addition of the correlated components is applied—that is, worst-case estimates of the measurement uncertainty are used.

3.3 EVALUATION OF CONCURRENT MEASUREMENT RESULTS

Concurrent measurement is an indirect measurement in which the values of several simultaneously measured heterogeneous physical quantities are obtained by solving equations that relate them to other physical quantities measured directly or indirectly. Concurrent measurements are a type of measurement of relationships.

Concurrent measurements of two or more heterogeneous quantities are performed to determine the relationship between them. They are based on known equations that represent the connections between the properties of objects.

The most widely used method for processing concurrent measurements is the least squares method (LSM). Its essence is as follows. In concurrent measurements, the unknown quantities subject to direct measurement are determined based on the results of measurements of other quantities that are functionally related to them:

$$Y = f(X_1, X_2, \ldots, X_m, A_1, A_2, \ldots, A_m, \ldots, Z_1, Z_2, \ldots, Z_m). \quad (3.48)$$

In practice, the dependence of Y on a single argument is usually determined—for example, when parameter X is measured: $Y = f(X)$.

The measured quantities in this system contain errors. Suppose we have a system of linear equations $a_k x_1 + b_k x_2 + c_k x_3 - y_k = 0$, where $k = 1, 2, 3, \ldots, n$ and x_1, x_2, x_3 are the unknown concurrent measured quantities. Since measurement errors always exist, the right-hand sides of the linear equations will not be zero. If we add a certain term δ_k, called the residual (the final error of the conditional equations), to the right-hand side, we obtain the following system of linear equations[5]:

$$\delta_k = a_k x_1 + b_k x_2 + c_k x_3 - y_k. \quad (3.49)$$

The problem reduces to finding such values $x_1, x_2, \ldots, x_n$ that, when substituted into the system of n conditional equations:

$$\left.\begin{aligned} a_1 x_1 + b_1 x_2 + c_1 x_3 - y_1 &= \delta_1; \\ a_2 x_1 + b_2 x_2 + c_2 x_3 - y_2 &= \delta_2; \\ \ldots\ldots\ldots\ldots\ldots\ldots\ldots\ldots\ldots \\ a_n x_1 + b_n x_2 + c_n x_3 - y_n &= \delta_n, \end{aligned}\right\} \quad (3.50)$$

provide the minimum value of the sum of the squares of deviations (residuals) of the experimental values from the calculated ones, that is:

$$\sum_{k=1}^{n} (a_k x_1 + b_k x_2 + c_k x_3 - y_k)^2 = \sum_{k=1}^{n} \delta_k^2 = Q = min. \quad (3.51)$$

During measurements with equal probability, a necessary condition for the minimum of the sum (3.51) is that the partial derivatives of the function (3.52) are equal to zero

$$\frac{\partial Q}{\partial x_1} = \frac{\partial Q}{\partial x_2} = \frac{\partial Q}{\partial x_3} = \ldots = \frac{\partial Q}{\partial x_m} = 0. \quad (3.52)$$

The conditions for the minimum, according to the Legendre principle for the system of equations, are as follows:

$$\left.\begin{aligned}\frac{\partial Q}{\partial x_1} &= 2\sum\nolimits_{k=1}^{n}(a_k x_1 + b_k x_2 + c_k x_3 - y_k)a_k = 0;\\ \frac{\partial Q}{\partial x_2} &= 2\sum\nolimits_{k=1}^{n}(a_k x_1 + b_k x_2 + c_k x_3 - y_k)b_k = 0;\\ \frac{\partial Q}{\partial x_3} &= 2\sum\nolimits_{k=1}^{n}(a_k x_1 + b_k x_2 + c_k x_3 - y_k)c_k = 0.\end{aligned}\right\} \tag{3.53}$$

As a result, from the set of equations obtained from the minimum condition, we have a system of m equations called the normal equations, whose coefficients depend on the coefficients of all n conditional equations:

$$\left.\begin{aligned}[aa]x_1 + [ab]x_2 + [ac]x_3 - [ay] = 0;\\ [ab]x_1 + [bb]x_2 + [bc]x_3 - [by] = 0;\\ [ac]x_1 + [bc]x_2 + [cc]x_3 - [cy] = 0,\end{aligned}\right\} \tag{3.54}$$

where

$$\left.\begin{aligned}[aa] &= a_1 a_1 + a_2 a_2 + \ldots + a_n a_n;\\ [ab] &= a_1 b_1 + a_2 b_2 + \ldots + a_n b_n;\\ [ac] &= a_1 c_1 + a_2 c_2 + \ldots + a_n c_n.\end{aligned}\right\} \tag{3.55}$$

Knowing the values of the constant coefficients of the normal system of equations, whose number equals the number of unknowns mmm, the results of concurrent measurements can be determined using known methods for solving linear equations, yielding estimates that are as close as possible to their true values.

To solve the system of linear normal equations, three common methods are used: the determinant method, the method of successive elimination of unknowns (Gaussian elimination), and the matrix method. When the number of normal equations is fewer than four, the average values of the concurrently measured quantities $x_1, x_2, \ldots, x_n$ are typically calculated using determinants:

$$\bar{x}_1 = \frac{D_{x1}}{D}, \quad \bar{x}_2 = \frac{D_{x2}}{D}, \quad \ldots, \quad \bar{x}_m = \frac{D_{xm}}{D}, \tag{3.56}$$

where the main determinant D for $m = 3$ is equal to

$$D = \begin{vmatrix}[aa] & [ab] & [ac]\\ [ba] & [bb] & [bc]\\ [ca] & [cb] & [cc]\end{vmatrix}, \tag{3.57}$$

and the determinants D_{x1}, D_{x2}, D_{x3} are calculated from the main determinant D by replacing the column of coefficients corresponding to the unknown x_m with the column of constants:

$$D_{x1} = \begin{vmatrix} [ay] & [ab] & [ac] \\ [by] & [bb] & [bc] \\ [cy] & [cb] & [cc] \end{vmatrix}; \tag{3.58}$$

$$D_{x2} = \begin{vmatrix} [aa] & [ay] & [ac] \\ [ba] & [by] & [bc] \\ [ca] & [cy] & [cc] \end{vmatrix}; \tag{3.59}$$

$$D_{x3} = \begin{vmatrix} [aa] & [ab] & [ay] \\ [ba] & [bb] & [by] \\ [ca] & [cb] & [cy] \end{vmatrix}. \tag{3.60}$$

The uncertainties of the quantities $\bar{x}_m$ obtained because of concurrent measurements are expressed by the following equations:

$$u(\bar{x}_1) = \sqrt{\frac{A_{11}}{D} \frac{\sum_{k=1}^{n} \delta_k^2}{(n-m)}}; \tag{3.61}$$

$$u(\bar{x}_2) = \sqrt{\frac{A_{22}}{D} \frac{\sum_{k=1}^{n} \delta_k^2}{(n-m)}}; \tag{3.62}$$

$$u(\bar{x}_3) = \sqrt{\frac{A_{33}}{D} \frac{\sum_{k=1}^{n} \delta_k^2}{(n-m)}}, \tag{3.63}$$

where A_{11}, A_{22}, A_{33} are the algebraic cofactors (adjugates) of the diagonal elements of the main determinant D, obtained by deleting from the determinant matrix the nth column and mth row corresponding to the measured quantity being estimated, followed by multiplication by $(-1)^{k+1}$. That is, to compute the adjugate A_{11}, the first column and the first row are removed, and the resulting determinant is calculated:

$$A_{11} = \begin{vmatrix} [bb] & [bc] \\ [cb] & [cc] \end{vmatrix}. \tag{3.64}$$

To determine the adjugate A_{22}, the second column and the second row are deleted from the determinant matrix, and the resulting determinant is evaluated:

$$A_{22} = \begin{vmatrix} [aa] & [ac] \\ [ca] & [cc] \end{vmatrix}. \tag{3.65}$$

To determine the adjugate A_{33}, the third column and the third row are removed from the matrix of the determinant, and the resulting minor determinant is evaluated:

$$A_{33} = \begin{vmatrix} [aa] & [ab] \\ [ba] & [bb] \end{vmatrix}. \tag{3.66}$$

The expanded uncertainty of determining the values of concurrently measured quantities $\bar{x}_m$ is calculated using the following equation:

$$U(\bar{x}_m) = k_p u(\bar{x}_m), \tag{3.67}$$

where k is the coverage factor, which is determined from Student's t-distribution with $(n - m)$ degrees of freedom and a given confidence level.

As the number m increases, the volume of the required work rises rapidly, and therefore, in practice, a polynomial of no higher than the third degree is typically used.

In the justification of the LSM in mathematical statistics, it is assumed that the measurement results satisfy the following conditions:

- the values of the arguments are known exactly;
- the measurement results contain only random errors, which are independent, have zero mean, and equal variances;
- the errors of the measured quantities are normally distributed.

Under such conditions, LSM provides unbiased estimates of the calculated unknown quantities. However, in practice, these conditions are not always fulfilled. In addition to the random components of measurement errors, there may also be unaccounted systematic components of uncertainty.

Confidence intervals for the true values of the measured quantities are determined based on the t-distribution with $(n - m)$ degrees of freedom, or from the normal distribution if the measurement results can be considered normally distributed.

An example of concurrent measurements is the determination of the parameters α and β in the equation expressing the dependence of resistance R on temperature t

$$R = R_{20} + \alpha(t - 20) + \beta(t - 20)^2, \tag{3.68}$$

where R_{20} is the resistance of the thermistor at 20°C; α, β are the temperature coefficients of resistance.

Equation (3.68) for the measured quantities α and β takes the form of a system of two equations corresponding to the temperature values t_1 and t_2:

$$\left.\begin{aligned} R_1 &= R_{20} + \alpha(t_1 - 20) + \beta(t_1 - 20)^2; \\ R_2 &= R_{20} + \alpha(t_2 - 20) + \beta(t_2 - 20)^2. \end{aligned}\right\} \tag{3.69}$$

Data processing in this case is carried out by analogy with indirect measurement. To do this, β is determined from the first equation of system (3.69)

$$\beta = \frac{\frac{R_1}{R_{20}} - 1 - \alpha(t_1 - 20)}{(t_1 - 20)^2}. \tag{3.70}$$

Equation (3.70) is substituted into the second equation of system (3.69), resulting in an equation for determining α:

$$\alpha = \frac{\left(\frac{R_2}{R_{20}} - 1\right)(t_1 - 20)^2 - \left(\frac{R_1}{R_{20}} - 1\right)(t_2 - 20)^2}{(t_1 - 20)(t_2 - 20)(t_1 - t_2)} \tag{3.71}$$

based on the measurement results of the quantities $t_1, t_2, R_1,$ and R_2.

If the measurement results of the quantities $t_1, t_2, R_1,$ and R_2 are given with expanded uncertainty, and the relative expanded uncertainties $\frac{U(x_i)}{x_i}$ are small, then interval analysis rules with linearization can be applied to obtain the expanded uncertainty:

$$U(\alpha) = \left|\frac{\partial\alpha}{\partial R_1}\right| U(R_1) + \left|\frac{\partial\alpha}{\partial R_2}\right| U(R_2) + \left|\frac{\partial\alpha}{\partial t_1}\right| U(t_1) + \left|\frac{\partial\alpha}{\partial t_2}\right| U(t_2). \tag{3.72}$$

If the result needs to be expressed in terms of combined uncertainty $u_c(\alpha)$, it can be calculated under the assumption of a uniform probability distribution using the equation:

$$u_c(\alpha) = \sqrt{\frac{\left[\frac{\partial\alpha}{\partial R_1} U(R_1)\right]^2 + \left[\frac{\partial\alpha}{\partial R_2} U(R_2)\right]^2 + \left[\frac{\partial\alpha}{\partial t_1} U(t_1)\right]^2 + \left[\frac{\partial\alpha}{\partial t_2} U(t_2)\right]^2}{3}}. \tag{3.73}$$

The expanded uncertainty $U(\beta)$ and the combined uncertainty $u_c(\beta)$ can be calculated using the error equation for the measurement of the quantity β:

$$\Delta(\beta) = \frac{\partial\beta}{\partial R_1}\Delta(R_1) + \frac{\partial\beta}{\partial\alpha}\Delta(\alpha) + \frac{\partial\beta}{\partial t_1}\Delta(t_1), \tag{3.74}$$

where $\Delta(\alpha) = \frac{\partial\alpha}{\partial R_1}\Delta(R_1) + \frac{\partial\alpha}{\partial R_2}\Delta(R_2) + \frac{\partial\alpha}{\partial t_1}\Delta(t_1) + \frac{\partial\alpha}{\partial t_2}\Delta(t_2)$.

The derivative is $\frac{\partial\beta}{\partial\alpha} = \frac{1}{t_1 - 20}$.

Substituting $\Delta(\alpha)$ and $\frac{\partial\beta}{\partial\alpha}$ into Equation (3.74), we obtain the error equation for the quantity β

$$\Delta(\beta) = \left(\frac{\partial\beta}{\partial R_1} - \frac{1}{(t_1 - 20)}\frac{\partial\alpha}{\partial R_1}\right)\Delta(R_1) - \frac{1}{(t_1 - 20)}\frac{\partial\alpha}{\partial R_2}\Delta(R_2) +$$

$$+ \left(\frac{\partial\beta}{\partial t_1} - \frac{1}{(t_1 - 20)}\frac{\partial\alpha}{\partial t_1}\right)\Delta(t_1) - \frac{1}{(t_1 - 20)}\frac{\partial\alpha}{\partial t_2}\Delta(t_2). \tag{3.75}$$

The expanded uncertainty of the quantity β is calculated using Equation (3.69) considering the rules of interval analysis:

$$U(\beta) = \left|\frac{\partial \beta}{\partial R_1} - \frac{1}{(t_1 - 20)}\frac{\partial \alpha}{\partial R_1}\right| U(R_1) + \left|\frac{1}{(t_1 - 20)}\frac{\partial \alpha}{\partial R_2}\right| U(R_2) +$$

$$+ \left|\frac{\partial \beta}{\partial t_1} - \frac{1}{(t_1 - 20)}\frac{\partial \alpha}{\partial t_1}\right| U(t_1) + \left|\frac{1}{(t_1 - 20)}\frac{\partial \alpha}{\partial t_2}\right| U(t_2). \tag{3.76}$$

The combined uncertainty of the quantity β is found from Equation (3.75) under the assumption of a uniform probability distribution of the uncertainties of the arguments:

$$u_c(\beta) = \frac{1}{\sqrt{3}}\left\{\left[\left(\frac{\partial \beta}{\partial R_1} - \frac{1}{(t_1 - 20)}\frac{\partial \alpha}{\partial R_1}\right) U(R_1)\right]^2 + \left[\frac{1}{(t_1 - 20)}\frac{\partial \alpha}{\partial R_2} U(R_2)\right]^2 +\right.$$

$$\left. + \left[\left(\frac{\partial \beta}{\partial t_1} - \frac{1}{(t_1 - 20)}\frac{\partial \alpha}{\partial t_1}\right) U(t_1)\right]^2 + \left(\frac{1}{(t_1 - 20)}\frac{\partial \alpha}{\partial t_2} U(t_2)\right)^2\right\}^{0,5}. \tag{3.77}$$

The measurement results of the quantities α and β, together with their combined uncertainties, are recorded in the form:

$$\alpha; \quad u_c(\alpha); \qquad \beta; \quad u_c(\beta). \tag{3.78}$$

When presenting expanded uncertainty, the measurement results are recorded as follows:

$$\alpha \pm U(\alpha); \quad \beta \pm U(\beta), \quad p. \tag{3.79}$$

3.4 EVALUATION OF COLLECTIVE MEASUREMENT RESULTS

A collective measurement is an indirect measurement in which the values of several simultaneously measured homogeneous quantities are obtained by solving a system of equations that relate various combinations of these quantities, measured either directly or indirectly.

Collective measurements are widely used in metrological practice, particularly for the calibration of measures or scales of measuring instruments. The processing of measurement results is performed using the LSM. However, the application of LSM to the processing of collective measurements has certain specific features. These are primarily due to the fact that the equations for the collective homogeneous quantities $y_1, y_2, \ldots, y_m$ take the following form:

$$x_i = \sum\nolimits_{j=1}^{m} C_{ij} y_j, \text{for i} = 1, 2, \ldots, n, \tag{3.80}$$

where y_j are the results of measuring the initial quantities; C_{ij} are known coefficients; x_i are the results of comparing various combinations of measures or scale divisions;

m is the number of quantity values to be determined; n is the number of combinations (equations).[6]

In addition, in many cases, additional exact conditions—so-called constraint equations—are introduced based on the physical conditions of the measurements to relate the quantities being determined. When constraint equations are available, it is possible to express some of the unknown quantities in terms of others.

During calibration, the coefficients C_{ij} take the following values:

1. 0—if Y_j does not participate in the ith measurement;
2. 1—if the sum of several quantities, including Y_j are measured;
3. -1—if the sum of several quantities is compared with Y_j.

If the number of equations equals the number of unknowns, the system (3.80) has a unique solution, and the true values of the measured quantities, along with their standard and expanded uncertainties, are determined using methods for processing indirect measurements. However, to reduce uncertainties, calibration is typically performed by comparing a greater number of combinations than the number of values of the quantities being determined. In this case, the evaluation of measurement results is carried out in the same manner as for concurrent measurements.

To solve the system of conditional equations, the LSM is usually applied. This method is derived from the principle of maximum likelihood and is optimal under the following conditions:

- The measurement results x contain independent random errors with zero expected values and equal variances.
- The errors are normally distributed.

When these conditions are met, the obtained estimates are unbiased and efficient. The system of equations with respect to the residuals can be written in the form:

$$\delta_i = \sum\nolimits_{j=1}^{m} C_{ij} y_i - x_i. \tag{3.81}$$

The sum of squares is equal to:

$$Q = \sum\nolimits_{i=1}^{n} \delta_i^2 = \sum\nolimits_{i=1}^{n} (C_{i1} y_1 + \ldots + C_{im} y_m - x_i)^2. \tag{3.82}$$

By differentiating expressions (3.82) with respect to the parameters y_j, the following system is obtained:

$$\left.\begin{aligned} \frac{\partial Q}{\partial y_1} &= 2\sum\nolimits_{i=1}^{n} (C_{i1} y_1 + \ldots + C_{im} y_m - x_i) C_{i1} \\ &\ldots \\ \frac{\partial Q}{\partial y_m} &= 2\sum\nolimits_{i=1}^{n} (C_{i1} y_1 + \ldots + C_{im} y_m - x_i) C_{im} \end{aligned}\right\}, \tag{3.83}$$

by further transforming and using Gauss notation, the normal system of equations for y_j is obtained:

$$\left.\begin{array}{l} [xC_1] = [C_1{}^2]y_1 + \ldots + [C_1C_m]y_m \\ \ldots \\ [xC_m] = [C_1C_m]y_1 + \ldots + [C_m{}^2]y_m \end{array}\right\}. \tag{3.84}$$

The solution of this system (3.84) by means of determinants is expressed as:

$$y_j = \frac{D_j}{D}, \tag{3.85}$$

where D is the main determinant of the system:

$$D = \begin{vmatrix} [C_1^2] & [C_1C_2] & \ldots & [C_1C_m] \\ [C_1C_2] & [C_2^2] & \ldots & [C_2C_m] \\ \ldots & \ldots & \ldots & \ldots \\ [C_1C_m] & [C_2C_m] & \ldots & [C_m^2] \end{vmatrix}, \tag{3.86}$$

and the determinants D_j are obtained from the main determinant by replacing the jth column with the column of constants:

$$D_j = \begin{vmatrix} [C_1^2] & \ldots & [YC_1] & \ldots & [C_1C_m] \\ [C_1C_2] & \ldots & [YC_{21}] & \ldots & [C_2C_m] \\ \ldots & \ldots & \ldots & \ldots & \ldots \\ [C_1C_m] & \ldots & [YC_{m1}] & \ldots & [C_m^2] \end{vmatrix}. \tag{3.87}$$

The experimental standard deviations y_j are determined using the following equation:

$$u(y_j) = \sqrt{\frac{A_{jj}}{D}\frac{\sum_{i=1}^{n}\delta_i^2}{n-m}}, \tag{3.88}$$

where A_{jj} are the algebraic cofactor of the main determinant, obtained by deleting the jth column and the jth row from it.

The residuals δ_i are determined using Equation (3.81).

The expanded uncertainties of the collective measurement results are determined from the expression:

$$U(y_j) = t_p(v)u(y_j), \tag{3.89}$$

where $t_p(v)$ is Student's coefficient for $v = n - m$ degrees of freedom.

This method of processing experimental data for collective measurements is appropriate for linear functions. In other cases, the processing of results becomes significantly more complex.

Notes

1 JCGM 100:2008, *Evaluation of measurement data—Guide to the expression of uncertainty in measurement* (Int. Organ. Stand., Geneva, ISBN 50, (2008): 134.
2 V. O. Podzharenko, O. M. Vasilevskyi, V. Y. Kucheruk, *Processing of measurement results based on the concept of uncertainty*, Vinnytsia: VNTU, (2008): 128.
3 C. Carmeli, T. Heinosaari, D. Reitzner, J. Schultz, A. Toigo, Quantum Incompatibility in Collective Measurements, *Mathematics*, Volume 4, Issue 3, 2016, 54.
4 O. M. Vasilevskyi, V. Y. Kucheruk, Y. T. Volodarskyi, *Uncertainty of measurement, control and test results* OLDI-PLUS, (2020): 352.
5 Xiaodong Li, Ruixin Su, Yilin Chen, Tianming Yang, Optimal policy for uncertainty estimation concurrent with decision making, *Cell Reports*, Volume 42, Issue 3, 2023, 112232.
6 Jan Roik, Karol Bartkiewicz, Antonín Černoch, Karel Lemr, Entanglement quantification from collective measurements processed by machine learning, *Physics Letters A*, Volume 446, 2022, 128270.

BIBLIOGRAPHY

Carmeli, C., T. Heinosaari, D. Reitzner, J. Schultz, and A. Toigo, Quantum incompatibility in collective measurements, *Mathematics*, *4*(3), 2016, 54. https://doi.org/10.3390/math4030054

JCGM 100:2008, *Evaluation of measurement data—Guide to the expression of uncertainty in measurement*, Int. Organ. Stand., Geneva, ISBN 50, (2008): 134.

Li, Xiaodong, Ruixin Su, Yilin Chen, and Tianming Yang, Optimal policy for uncertainty estimation concurrent with decision making, *Cell Reports*, 42, 3, 2023, 112232. https://doi.org/10.1016/j.celrep.2023.112232

Podzharenko, V. O., O. M. Vasilevskyi, and V. Y. Kucheruk, *Processing of measurement results based on the concept of uncertainty*, Vinnytsia: VNTU, (2008): 128.

Roik, Jan, Karol Bartkiewicz, Antonín Černoch, and Karel Lemr, Entanglement quantification from collective measurements processed by machine learning, *Physics Letters A*, 446, 2022, 128270. https://doi.org/10.1016/j.physleta.2022.128270

Vasilevskyi, O. M., V. Y. Kucheruk, and Y. T. Volodarskyi, *Uncertainty of measurement, control and test results: A textbook*, Kherson: OLDI-PLUS, (2020): 352, ISBN: 978-966-289-374-8.

4 Practical Recommendations and Examples of Evaluating Uncertainty Components

4.1 COMPARATIVE ANALYSIS OF TWO APPROACHES TO EXPRESSING MEASUREMENT ACCURACY CHARACTERISTICS

The purpose of measurement is to obtain an estimate of the true value of the measurand. The concept of measurement error, defined as the difference between the measurement result and the true (actual) value of the measurand, is used to describe measurement accuracy in normative documents on metrology within the field of ensuring the uniformity of measurements. When referring to the evaluation of errors in metrological practice, what is meant is the evaluation of its characteristics (Figure 4.1).

According to standards and normative documents,[1,2] the concept of measurement uncertainty is employed to represent measurement accuracy. Measurement uncertainty is interpreted as incomplete knowledge of the value of the measurand. To quantify this incompleteness, a probability distribution of the possible (justifiably assigned) values of the measurand is introduced. Consequently, a parameter of this distribution (the uncertainty) provides a quantitative characterization of the accuracy of the measurement result.

The procedure for both approaches—estimation of error characteristics and evaluation of measurement uncertainty—follows the same sequence of operations:

- Analysis of the measurement equation.
- Identification of all sources of measurement error (uncertainty) and their quantitative assessment.
- Application of corrections for systematic errors (effects) that can be eliminated.

Methods for calculating uncertainty, as well as methods for estimating error characteristics, are borrowed from mathematical statistics, but different interpretations of the probability distribution law of random variables are used. In addition to those set out in normative documents on the expression of measurement uncertainty, other methods are also widely used in practice.

The differences between the two approaches manifest themselves in the interpretation of uncertainty and error characteristics, which are based on different

DOI: 10.1201/9781003742708-4

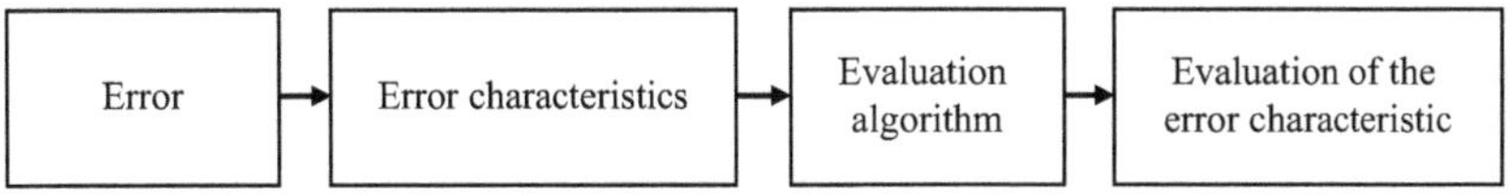

FIGURE 4.1 Approach used in the evaluation of quantities characterizing measurement errors.

interpretations of probability: frequentist and subjective. For example, the confidence limits of the error $(y - \Delta_p; y + \Delta_p)$ (offset from the measurement result) cover the true value of the measured quantity with a given confidence probability (frequentist interpretation of probability). At the same time, a similar interval $(y - U_p; y + U_p)$ is interpreted in standards on the expression of measurement uncertainty as an interval that contains a given fraction of the distribution of values that could reasonably be attributed to the measured quantity (subjective interpretation of probability).

In the general case, there is no one-to-one correspondence between random errors and uncertainties calculated by Type A, as well as between residual systematic errors and uncertainties calculated by Type B. The division into systematic and random errors is determined by the nature of their origin and manifestation during the measurement experiment, while the division into uncertainties calculated by Type A and Type B is determined by the methods of their evaluation.

The procedure for estimating the error characteristics of a measurement result is presented in Table 4.1.[3,4]

The procedure for evaluating measurement uncertainty is given in Table 4.2.

In comparing the estimates of error characteristics and measurement uncertainties, the following correspondence scheme is suggested:

RMS characterizing the random error	↔	Standard uncertainty of Type A
RMS characterizing the residual systematic error	↔	Standard uncertainty of Type B
RMS characterizing the total error	↔	Combined uncertainty
Confidence limits of the error	↔	Expanded uncertainty

TABLE 4.1

Procedure for Evaluating Error Characteristics

Error	$\Delta y = y_m - y_t \Leftrightarrow y_m = y_t + \Delta y$
Error model	Δy is a random variable with a probability density function $p(x, M, \sigma^2)$, where M is the mathematical expectation and σ^2 is the variance.
Error characteristics	S is standard deviation; Θ is limits of residual systematic error; Δ_p is confidence limits
Input data for assessing error characteristics	1. Model of the investigated object. 2. Experimental data x_{iq}, where $q = 1, \ldots, n; i = 1, \ldots, m$. 3. Information on the distribution laws. 4. Information on error sources, their nature, and component characteristics $(S(x_i), \Theta_i)$. 5. Standard reference information and other reference materials.

(Continued)

Table 4.1 (Continued)

Methods for evaluating error characteristics:	
1. Random errors	$S(x_{iq}) = \sqrt{\frac{\sum_{q=1}^{n_i}(x_{iq}-\bar{x}_i)^2}{n_i-1}}$; $S(\bar{x}_i) = \sqrt{\frac{\sum_{q=1}^{n_i}(x_{iq}-\bar{x}_i)^2}{(n_i-1)n_i}}$; $S_{\sum} = \sqrt{\sum_{i=1}^{m}\left(\frac{\partial f}{\partial x_i}\right)^2 s^2(\bar{x}_i)}$.
2. Residual systematic errors	$\Theta(p) = k\sqrt{\sum_{i=1}^{m}\left(\frac{\partial f}{\partial x_i}\right)^2 \Theta_i^2}$, where $k = 1.1$ for $p = 0.95; k = 1.4$ for $p = 0.99$ and $m > 4$.
3. Total error	$\Delta_p = \frac{t_p(f_{\text{eff}})S_{\sum} + \Theta(p)}{S_{\sum} + \sqrt{\sum_{i=1}^{m}\left(\frac{\partial f}{\partial x_i}\right)^2 \frac{\Theta_i^2}{3}}}\sqrt{S_{\sum}^2 + \sum_{i=1}^{m}\left(\frac{\partial f}{\partial x_i}\right)^2 \frac{\Theta_i^2}{3}}$, when the inequality $0.8 \le \frac{\Theta(p)}{S_{\sum}} \le 8.0$ holds; $\Delta_p = t_p(f_{\text{eff}})S_{\sum}$ for $\frac{\Theta(p)}{S_{\sum}} < 0.8$ the systematic part of the error is disregarded; $\Delta_p = \Theta(p)$ for $\frac{\Theta(p)}{S_{\sum}} > 8.0$ the random part of the error is disregarded; $f_{\text{eff}} = \frac{\left(\sum_{i=1}^{m}\left(\frac{\partial f}{\partial x_i}\right)^2 s^2(\bar{x}_i)\right)^2 - \frac{2\sum_{i=1}^{m}\left(\frac{\partial f}{\partial x_i}\right)^4 s^4(\bar{x}_i)}{m+1}}{\frac{\sum_{i=1}^{m}\left(\frac{\partial f}{\partial x_i}\right)^4 s^4(\bar{x}_i)}{m+1}}$.
Representation of error characteristics	$\Theta(p), S_{\sum}, n, f_{\text{eff}}, \Delta_p$.
Interpretation of the results obtained	The interval $(-\Delta_p; +\Delta_p)$ includes the measurement error with probability p, which is equivalent to saying that the interval $(y - \Delta_p; y + \Delta_p)$ contains the true value of the measured quantity with probability p.

TABLE 4.2
Procedure for Evaluating Measurement Uncertainty

Model of uncertainty (representation of our knowledge of the measured quantity's value)	ε is a random variable with probability density function $p(x, y, u^2)$, where y is the expected value and u^2 is the variance.		
Measurement uncertainty (quantitative measure)	Standard uncertainty u	Combined uncertainty $u_c = \sqrt{\sum_{i=1}^{m} u_i^2}$	Expanded uncertainty $U_p = ku_c$
Input data for assessing measurement uncertainty	1. Model of the investigated object. 2. Experimental data x_{iq}, where $q = 1, \ldots, n, i = 1, \ldots, m$. 3. Information on the distribution laws. 4. Information on the sources of uncertainty and on the values of uncertainty. 5. Standard reference data and other reference materials.		

(Continued)

Table 4.2 (Continued)

Methods for evaluating measurement uncertainty:	
1. Type A evaluation	$u_{A,i} = \sqrt{\frac{\sum_{q=1}^{n_i}\left(x_{iq}-\bar{x}_i\right)^2}{n_i-1}}$; $u_A(\bar{x}_i) = \sqrt{\frac{\sum_{q=1}^{n_i}\left(x_{iq}-\bar{x}_i\right)^2}{(n_i-1)n_i}}$.
2. Type B evaluation	$u_B(x_i) = \frac{\alpha_+ - \alpha_-}{\sqrt{12}}$.
3. Expanded uncertainty interval	$U_p = t_p(\nu_{eff})\sqrt{\sum_{i=1}^{m}\left(\frac{\partial f}{\partial x_i}\right)^2 u^2(x_i)}$, where $\nu_{eff} = \frac{u_c^4}{\sum_{i=1}^{m}\left(\frac{\partial f}{\partial x_i}\right)^4 \frac{u^4(x_i)}{\nu_i}}$, where $\nu_i = n_i - 1$ for Type A uncertainties, and $\nu_i = \infty$ for Type B uncertainties. In most practical cases, it can be assumed that: $U_{0.95} = 2u_c, U_{0.99} = 3u_c$ for a normal distribution; $U_{0.95} = 1.65u_c, U_{0.99} = 1.71u_c$ for a uniform distribution.
Representation of uncertainty	u_c, U_p, k, u_i, ν_i.
Interpretation of the results obtained	The interval $(y - U_p; y + U_p)$ encompasses a large fraction (p) of the distribution of values that can be justifiably assigned to the measured quantity.

In the absence of information for evaluating uncertainty $\hat{u}$, it may be estimated from the assessed error characteristics using the schemes shown in Figures 4.2 and 4.3.

The schemes in Figures 4.2 and 4.3 represent two distinct approaches to presenting measurement results, as prescribed in metrological standards.

Evaluation of Type A and Type B uncertainties separately is impossible when only the confidence limit Δ_p is known.

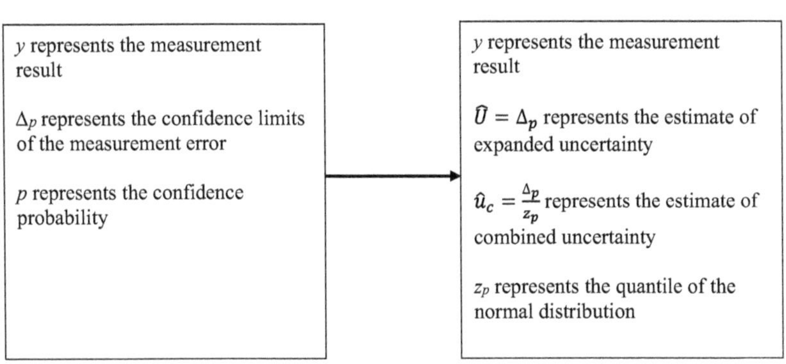

FIGURE 4.2 Procedure for assessing uncertainty using the confidence limits of measurement errors.

y represents the measurement result

S represents the RMS of the random error of the measurement result

$\Theta(p)$ represents the residual systematic error

m represents the number of measured input quantities

f_{eff} represents the estimated effective number of degrees of freedom

$f_{eff} = n - 1$ for direct measurements, where n is the number of measurements

→

y represents the measurement result

$\hat{u}_A = S$ represents the estimate of the Type A standard uncertainty

$\hat{u}_B = \frac{\Theta(p)}{k\sqrt{3}}$ represents the estimate of the Type B standard uncertainty, in this case:
$k = 1.1$ for $p = 0.95$;
$k = 1.4$ for $p = 0.99$ and $m > 4$

$\hat{u}_c = \sqrt{\hat{u}_A{}^2 + \hat{u}_B{}^2}$ represents the estimate of combined uncertainty

$\hat{\nu}_{eff} = f_{eff}\left(1 + \frac{\hat{u}_B{}^2}{\hat{u}_A{}^2}\right)^2$ represents the estimated effective number of degrees of freedom

$\hat{U} = t_p(\hat{\nu}_{eff})\hat{u}_c$ represents the estimate of expanded uncertainty

FIGURE 4.3 Procedure for assessing uncertainty using the RMS of the random measurement error and the confidence limits of residual systematic errors.

4.2 METHOD FOR CONVERTING ERROR CHARACTERISTICS TO MEASUREMENT UNCERTAINTY CHARACTERISTICS

When comparing measurement results and the assessments of their reliability obtained by different methods, the task arises of converting error characteristics into uncertainty characteristics and vice versa. Therefore, we consider a methodology for such conversion and the minimal set of parameters required to perform it.

The conversion method shown in Figure 4.3 allows the calculation of expanded uncertainty using known characteristics of random and systematic errors for equal confidence levels and confidence probability. However, the confidence limits of the errors of primary and secondary standards must be evaluated for a confidence probability of 0.99. In such cases, it is preferable to use formulas for mutual conversion between the confidence error and expanded uncertainty.

The input data for converting error characteristics into uncertainty characteristics are: the confidence limits of the total measurement error Δ_P; the confidence probability p; the ratio of the confidence limits $\gamma = \Theta(p)/S_{\sum}$ of the residual systematic error $\Theta(p)$ to the root mean square (RMS) of the random error $S_{\sum}$; the number of repeated measurements n; and the number of residual summable systematic errors m.

To convert the confidence limits of errors from direct single measurements, known with probabilities of 0.95 or 0.99, into expanded uncertainty with a confidence level of 0.95, the following expressions should be used:

- For the error with the probability $p = 0.95$:

$$U_{0.95} = \begin{cases} \Delta_{0.95}\sqrt{1+\dfrac{\gamma^2}{3.63}} & \text{for } \gamma < 0.8, \\ \dfrac{2\Delta_{0.95}}{K[\gamma+2]}\sqrt{1+\dfrac{\gamma^2}{3.63}} & \text{for } 0.8 \le \gamma \le 8, \\ 2\Delta_{0.95}\sqrt{\dfrac{1}{\gamma^2}+\dfrac{1}{3.63}} & \text{for } \gamma > 8. \end{cases} \tag{4.1}$$

- For the error with the probability $p = 0.99$:

$$U_{0.95} = \begin{cases} \dfrac{\Delta_{0.99}}{1.3}\sqrt{1+\dfrac{\gamma^2}{3d^2}}, & \text{for } \gamma < 0.8, \\ \dfrac{2\Delta_{0.99}}{K[\gamma+2.6]}\sqrt{1+\dfrac{\gamma^2}{3d^2}}, & \text{for } 0.8 \le \gamma \le 8, \\ \dfrac{2\Delta_{0.99}}{d}\sqrt{\dfrac{1}{\gamma^2}+\dfrac{1}{3d^2}}, & \text{for } \gamma > 8, \end{cases} \tag{4.2}$$

where d is the correction factor, which for $p = 0.99$ equals 1.2 for $m = 2$; 1.3 for $m = 3$; 1.4 for $m = 4$; and 1.45 for $m >$; K is the coefficient, the value of which depending on p and γ is given in Table 4.3.

When converting the confidence limits of the error of direct repeated measurements, known with a probability of 0.95 or 0.99, into the expanded uncertainty with a confidence level of 0.95, the following formulas should be used:

TABLE 4.3
Values of Coefficient K for Different p and γ

γ	0.8	1	2	3	4	5	6	7	8
$K_{0.95}$	0.76	0.74	0.71	0.73	0.76	0.78	0.79	0.80	0.81
$K_{0.99}$	0.84	0.82	0.80	0.81	0.82	0.83	0.83	0.84	0.85

- For the error with the probability $p = 0.95$:

$$U_{0.95} = \begin{cases} \dfrac{\Delta_{0.95} t_{0.95}(\nu_{\text{eff}})}{t_{0.95}(n-1)} \sqrt{1 + \dfrac{\gamma^2}{3.63}}, & \text{for } \gamma < 0.8, \\ \dfrac{\Delta_{0.95} t_{0.95}(\nu_{\text{eff}})}{K[\gamma + t_{0.95}(n-1)]} \sqrt{1 + \dfrac{\gamma^2}{3.63}}, & \text{for } 0.8 \le \gamma \le 8, \\ \Delta_{0.95} t_{0.95}(\nu_{\text{eff}}) \sqrt{\dfrac{1}{\gamma^2} + \dfrac{1}{3.63}}, & \text{for } \gamma > 8. \end{cases} \tag{4.3}$$

- For the error with the probability $p = 0.99$:

$$U_{0.95} = \begin{cases} \dfrac{\Delta_{0.99} t_{0.95}(\nu_{\text{eff}})}{t_{0.99}(n-1)} \sqrt{1 + \dfrac{\gamma^2}{5.88}}, & \text{for } \gamma < 0.8, \\ \dfrac{\Delta_{0.99} t_{0.95}(\nu_{\text{eff}})}{K[\gamma + t_{0.99}(n-1)]} \sqrt{1 + \dfrac{\gamma^2}{5.88}}, & \text{for } 0.8 \le \gamma \le 8, \\ \Delta_{0.99} t_{0.95}(\nu_{\text{eff}}) \sqrt{\dfrac{1}{\gamma^2} + \dfrac{1}{5.88}}, & \text{for } \gamma > 8, \end{cases} \tag{4.4}$$

where $t_{0.95}(\nu_{\text{eff}})$ and $t_p(n-1)$ are Student's coefficients, corresponding to the probability p, the number of degrees of freedom $(n-1)$, and the effective number of degrees of freedom ν_{eff}, which is determined by the equation:

$$\nu_{\text{eff}} = (n-1)\left[1 + \frac{\gamma^2}{3d^2}\right]^2. \tag{4.5}$$

According to standards, when presenting a measurement result, either the expanded uncertainty or the combined standard uncertainty can be indicated as a measure of uncertainty. When presenting a measurement result with expanded uncertainty, the following characteristics are specified:

- the measurement result y;
- the expanded uncertainty U_p;
- the coverage factor k_p;
- the confidence level p;
- the number of Type B standard uncertainties (for $p = 0.99$).

In this case, the confidence limits of the errors of direct single measurements can be obtained from Equations (4.2) and (4.4).

In addition to the characteristics listed above, for the calculation of errors of repeated measurements, it is necessary to know:

- the number of observations n;
- the effective number of degrees of freedom ν_{eff}.

In this way, one can obtain:

- an estimate of the RMS deviation characterizing the total error

$$S_{\sum} = \frac{U_p}{k_p} = u_c. \tag{4.6}$$

- an estimate of the RMS of the random error

$$S = u_A = S_{\sum} \sqrt[4]{\frac{n-1}{\nu_{\text{eff}}}}. \tag{4.7}$$

- an estimate of the RMS characterizing the residual systematic error

$$S_{\Theta} = u_B = \sqrt{S_{\sum}^2 - S^2}. \tag{4.8}$$

- an estimate of the confidence limits of the residual systematic error

$$\Theta(P) = k_p S_{\Theta} \sqrt{3}. \tag{4.9}$$

- an estimate of the confidence limits of the total error

$$\Delta_P = \frac{t_P(n-1)S + \Theta(P)}{S + S_{\Theta}} S_{\sum}. \tag{4.10}$$

If the effective number of degrees of freedom is not available, it can be obtained using the coverage factor k_p from the t-distribution table.

4.3 METHOD FOR CONVERTING UNCERTAINTY CHARACTERISTICS INTO ERROR CHARACTERISTICS

According to standards, the uncertainty of a measurement result may be expressed either as a combined standard uncertainty or as an expanded uncertainty, particularly when the measurements concern human health or the environment.

If the uncertainty of a measurement result has been expressed as a combined standard uncertainty, it can be converted into the following error characteristics:

- the estimate of Type A standard uncertainty $u_A(\bar{x})$ is converted into the estimate of the RMS of the random error $S(\bar{x})$, i.e., $S(\bar{x}) = u_A(\bar{x})$;
- the estimate of Type B standard uncertainty u_B is converted into the estimate of the RMS of the residual systematic error $\Theta(p)$, i.e., $\Theta(p) = u_B$;

- the estimate of the combined uncertainty $u_c(y)$ is converted into the estimate of the RMS characterizing the total error S_Σ, i.e., $S_\Sigma = u_c(y)$.

If the measurement result is expressed as an expanded uncertainty, the following conversions from uncertainty characteristics to error characteristics are performed:

- the expanded uncertainty U_p is converted into the estimate of the confidence limits of the error Δ_p, i.e., $\Delta_p = U_p = k_p u_c(y)$;
- the combined uncertainty $u_c(y)$ is converted into the estimate of the RMS characterizing the total error S_Σ, i.e., $S_\Sigma = U_p/k_p$.

4.4 EXAMPLES OF EVALUATING MEASUREMENT, CONTROL, AND TESTING UNCERTAINTIES

4.4.1 Evaluation of Measurement Uncertainty during Electric Current Measurement

The circuit for measuring electric current using a voltmeter is shown in Figure 4.4.

The measurement equation is given by

$$I = f(V, R) = \frac{V}{R}, \tag{4.11}$$

where I is the electric current, U is the voltage, and R is the shunt resistance.

As a result of the voltage measurement, the following values U in millivolts were obtained: 100; 100.68; 100.83; 100.79; 100.64; 100.63; 100.94; 100.6; 100.68; 100.76; 100.64 at a temperature of $t = (23 \pm 0.05)$°C.[5]

Based on the obtained values, the mean voltage is calculated using the following equation:

$$\bar{V} = \frac{1}{n}\sum_{i=1}^{n} V_i = 100.72 \text{ mV}. \tag{4.12}$$

The value of the shunt resistance was determined during calibration for $I = 10$ A and $t = 23$°C, and is equal to $R_0 = 0.010088\ \Omega$.

The result of the current measurement is obtained using the following equation:

$$I = \frac{\bar{V}}{R_0} = 9.984 \text{ A}. \tag{4.13}$$

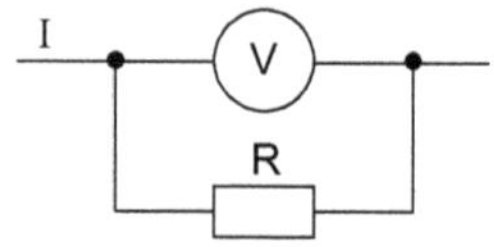

FIGURE 4.4 Electric current measurement circuit.

Next, the RMS deviation of the arithmetic mean voltage $S(\bar{V})$ is determined, which characterizes the random component of the error in voltage measurements, using the following equation:

$$S(\bar{V}) = \sqrt{\frac{\sum_{i=1}^{n}(V_i - \bar{V})^2}{n(n-1)}} = 3.4 \cdot 10^{-2}\ \text{mV}. \tag{4.14}$$

The relative value of the RMS deviation is calculated as follows:

$$\tilde{S}(\bar{V}) = \frac{S(\bar{V})}{\bar{V}} = 0.034\%. \tag{4.15}$$

The limits of the residual systematic error of the voltmeter were determined during its calibration and are expressed as follows:

$$\theta_V = \pm\left(3 \cdot 10^{-4} V + 0.02\right)\text{mV}. \tag{4.16}$$

Then, for $V = \bar{V}$, we obtain:

$$\theta_V = \pm 5.0 \cdot 10^{-2}\ \text{mV}$$

$$\tilde{\theta}_V = 0.05\%.$$

The limits of the residual systematic error of the shunt resistance, determined during its calibration, are equal to $\tilde{\theta}_R = 0.07\%$.

Then, for $R = R_0$, we obtain: $\theta_R = \pm 7 \cdot 10^{-4} \cdot R_0 = \pm 7.1 \cdot 10^{-6}\ \Omega$.

The limits of the residual systematic component of the shunt resistance error, caused by the error in temperature measurement, are determined using the following equation:

$$R = R_0[1 + \alpha(t - t_0)], \tag{4.17}$$

where R_0 is the resistance value at $t = t_0$ ($t_0 = 23°\text{C}; R_0 = 0.010088\ \Omega$) and α is the temperature coefficient $\left(\alpha = 6 \cdot 10^{-6} K^{-1}\right)$.

If the temperature measurement limits are Δt, then the limits of the corresponding component of the resistance error are $\theta_{R,t} = \pm \alpha \Delta t \cdot R$.

For $\Delta t = \pm 0.05°\text{C}$, we obtain: $\theta_{R,t} = \pm 3 \cdot 10^{-9}\ \Omega$, $\tilde{\theta}_{R,t} = 3 \cdot 10^{-5}\%$.

This component of the error can be neglected, as its numerical value is small compared to the other components.

To combine the residual systematic errors, it is assumed that they follow a uniform distribution within the limits θ_V and θ_R. Then, the RMS of the total residual systematic component of the current measurement error is determined using the following equation:

$$S_\theta = \sqrt{\left(\frac{\partial f}{\partial V}\right)^2 \frac{\theta_V^2}{3} + \left(\frac{\partial f}{\partial R}\right)^2 \frac{\theta_R^2}{3}}, \tag{4.18}$$

where $\frac{\partial f}{\partial V} = \frac{1}{R}$ and $\frac{\partial f}{\partial R} = -\frac{V}{R^2}$ are the sensitivity coefficients.

Then, we obtain:

$$S_\theta = \sqrt{\left(\frac{1}{R}\right)^2 \frac{\theta_V^2}{3} + \left(\frac{\bar{V}}{R^2}\right)^2 \frac{\theta_R^2}{3}} = 5 \cdot 10^{-3}\ \text{A}.$$

$$\tilde{S}_\theta = 0.05\%.$$

The confidence limits of the total residual systematic component of the current measurement error, for a probability $p = 0.95$, are estimated using the following equation:

$$\theta(0.95) = \pm 1.1\sqrt{\left(\frac{1}{R}\right)^2 \theta_V^2 + \left(\frac{-V}{R^2}\right)^2 \cdot \theta_R^2} = \pm 9.5 \cdot 10^{-3}\text{A}, \tag{4.19}$$

$$\tilde{\theta}_{(0.95)} = 0.95\%.$$

The RMS of the random component of the current measurement error is determined using the following equation:

$$S = \frac{\partial f}{\partial V} \cdot S(\bar{V}) = 3.4 \cdot 10^{-3}\text{A}, \tag{4.20}$$

$$\tilde{S} = 0.034\%.$$

The RMS of the total error of the current measurement result is calculated using the following equation:

$$S_\Sigma = \sqrt{S^2 + S_\theta^2} = 6 \cdot 10^{-3}\ \text{A}, \tag{4.21}$$

$$\tilde{S}_\Sigma = 0.06\%.$$

The confidence limits of the current measurement error, for $p = 0.95$ and the effective number of degrees of freedom $\nu_{\text{eff}} = n - 1 = 9$, are calculated using the following equation:

$$\Delta_{0.95} = \frac{t_{0.95}(9)S + \theta(0.95)}{S + S_\theta} \cdot S_\Sigma = 0.012\text{A}, \tag{4.22}$$

$$\tilde{\Delta}_{0.95} = 0.012\%.$$

Next, we use the approach based on the measurement uncertainty concept to evaluate the results of electric current measurements.

According to Type A, the standard uncertainty arising from a source of uncertainty with a random nature is calculated. The standard uncertainty of the voltage, due to a source of uncertainty with a random nature, is determined using the following equation:

$$u_A(\bar{V}) = \sqrt{\frac{\sum_{i=1}^{n}(V_i - \bar{V})^2}{n(n-1)}}, \tag{4.23}$$

$$u_A(\bar{V}) = 3.4 \cdot 10^{-2}\ \text{mV}, \tilde{u}_A = 0.034\%.$$

The standard uncertainty of the current, arising from a source of uncertainty with a random nature, is determined using the following equation:

$$u_A = \frac{\partial f}{\partial V} \cdot u_A(\bar{V}) = 3.4 \cdot 10^{-3}\text{A}, \tag{4.24}$$

$$\tilde{u}_A = 0.034\%.$$

According to Type B, the standard uncertainties arising from sources of uncertainty with a systematic nature are calculated. The distribution within the limits is assumed to be uniform.

The limits of the systematic deviation in voltage measurements, determined during the voltmeter calibration, are $3 \cdot 10^{-4} \cdot \bar{V} + 0.02$. The corresponding standard uncertainty is then calculated using the following equation:

$$u_{B,V} = \frac{3 \cdot 10^{-4} \cdot \bar{V} + 0.02}{\sqrt{3}} = 2.9 \cdot 10^{-2}\ \text{mV}, \tag{4.25}$$

$$\tilde{u}_{B,V} = 0.029\%.$$

The limits within which the shunt resistance lies, determined during the shunt calibration, are $7 \cdot 10^{-4} \cdot R$. Then, for $R = R_0$, the corresponding standard uncertainty is calculated using the following equation:

$$u_{B,R} = \frac{7 \cdot 10^{-4} \cdot R_0}{\sqrt{3}} = 4 \cdot 10^{-4}\ \Omega, \tag{4.26}$$

$$\tilde{u}_{B,R} = 0.04\%.$$

The limits of variation of the shunt resistance value, caused by a change in temperature, are $\alpha \cdot \Delta t \cdot R_0$. The corresponding standard uncertainty is obtained using the following equation:

$$u_{B,t} = \frac{\alpha \cdot \Delta t \cdot R_0}{\sqrt{3}} = 1.7 \cdot 10^{-9}\ \Omega, \tag{4.27}$$

$$\tilde{u}_{B,t} = 1.7 \cdot 10^{-5}\%.$$

Subsequently, this component of uncertainty (4.27) can be neglected.

The combined standard uncertainty, calculated according to Type B, is determined using the following equation:

$$u_B = \sqrt{\left(\frac{\partial f}{\partial V}\right)^2 U_{B,V}^2 + \left(\frac{\partial f}{\partial R}\right)^2 U_{B,R}^2} = 5 \cdot 10^{-3}\ \text{A}, \tag{4.28}$$

$$\tilde{u}_B = 0.05\%.$$

The combined uncertainty of the electric current measurement result is calculated as follows:

$$u_c = \sqrt{u_A^2 + u_B^2} = 6 \cdot 10^{-3}\ \text{A}, \tag{4.29}$$

$$\tilde{u}_c = 0.06\%.$$

The effective number of degrees of freedom is determined using the following equation:

$$\nu_{\text{eff}} = \frac{u_c^4}{\frac{\left(\frac{1}{R}\cdot u_A\right)^4}{n-1} + \frac{\left(\frac{1}{R}\cdot u_{B,V}\right)^4}{\infty} + \frac{\left(\frac{V}{R^2}\cdot u_{B,R}\right)}{\infty}} = 87. \tag{4.30}$$

The coverage factor k for this effective number of degrees of freedom is $k = t_{0.95}(\nu_{\text{eff}}) = t_{0.95}(87) = 1.96$.

The expanded uncertainty is determined using the following equation:

$$U_{0.95} = k \cdot u_c = 0.012\ \text{A}, \tag{4.31}$$

$$\tilde{U}_{0.95} = 0.12\%.$$

Thus, the measurement result can be expressed as follows:

$$I = (9.984 \pm 0.012)A, \text{at}\ p = 0.95.$$

4.4.2 Evaluation of the Measurement Uncertainty in Repeated Measurements of the Frequency of a Sinusoidal Signal

Repeated measurements were performed of the frequency of a sinusoidal signal in a general-purpose electrical power network. The frequency meter readings f (Hz) is

50.8; 49.7; 48.2; 48.8; 48.5; 49.3; 49.9; 50.4; 56.1; 51.1; 49.4; 51.6; 51.9; 50.5; 50.8; 49.9; 50.2; 50.4; 50.1; 50.3; 50.1.

It is required to evaluate the measurement uncertainty of the frequency at a confidence level $p = 0.95$.

A specification of these measurements was prepared:

- the measurements were carried out under laboratory conditions at an ambient air temperature of +23°C;
- the accuracy class of the frequency measuring instrument (MI) is 0.25;
- the temperature coefficient of frequency variation of the quartz resonator is not more than $\pm 1.5 \cdot 10^{-6}$ per 1°C.

The best estimate of the measurand, obtained from a series of repeated direct measurements of equal precision, which reduces the effect of random error in each observation, is the arithmetic mean:

$$\bar{f} = \frac{\sum_{k=1}^{20} f_k}{20} = 50.38 \text{ Hz}. \tag{4.32}$$

The standard deviation (SD) of the results from the arithmetic mean (Type A standard uncertainty) is

$$u_A = \sqrt{\frac{\sum_{k=1}^{n} (f_k - \bar{f})^2}{n - 1}} = 1.6 \text{ Hz}. \tag{4.33}$$

We will check the extreme values of the measurement results for outliers. For this purpose, we calculate the ratio:

$$s_1 = \frac{\bar{f} - f_{\min}}{u_A} = 1.36; \tag{4.34}$$

$$s_2 = \frac{f_{\max} - \bar{f}}{u_A} = 3.57. \tag{4.35}$$

Since the confidence level is taken as $p = 0.95$, the parameter α is calculated as $\alpha = 1 - p = 1 - 0.95 = 0.05$. According to this parameter and the number of measurements $n = 20$, from Table 3.1 we select the permissible value of the normalized deviation for the extreme values of the series, which is equal to $s_{crit} = 2.56$. By comparing the selected permissible value s_{crit} with the calculated values s_1 and s_2, the following conclusion can be drawn. Since the calculated parameter s_1 is less than the permissible value s_{crit}, the hypothesis regarding the presence of outliers in the observation series is rejected. This means that the smallest measurement result is not abnormal. When comparing the permissible value s_{crit} with the parameter for the largest observation s_2, it is found that $s_2 > s_{\text{crit}}$. This indicates that the maximum value in the observation series is an outlier. Outliers should be excluded from the measurement results. Therefore, the maximum frequency value of 56.1 Hz is discarded from the dataset, and the mean frequency and SD are recalculated:

$$\bar{f} = \frac{\sum_{k=1}^{19} f_k}{19} = 50.1 \text{ Hz};$$

$$u_{A2} = \sqrt{\frac{\sum_{k=1}^{n} (f_k - \bar{f})^2}{n-1}} = 0.95 \text{ Hz}.$$

After this step, the extreme values of the observation series are again tested for the presence of outliers:

$$s_1 = \frac{\bar{f} - f_{\min}}{u_{A2}} = 1.99;$$

$$s_2 = \frac{f_{\max} - \bar{f}}{u_{A2}} = 1.9.$$

For the given value of $\alpha = 0.05$ and the number of measurements $n = 19$, the new critical value of the normalized deviation for the extreme members of the series is taken from Table 3.1. Comparing this critical value $s_{\text{crit}} = 2.53$ with the newly calculated values s_1 and s_2, it is seen that both are smaller than the critical value. This means that the extreme members of the measurement results series do not contain outliers.

Therefore, since the outliers have been removed from the results of the repeated measurements, the standard uncertainty of the mean value of Type A can be evaluated using the equation

$$u_A(\bar{f}) = \sqrt{\frac{\sum_{k=1}^{n} (f_k - \bar{f})^2}{n(n-1)}} = \frac{u_{A2}}{\sqrt{n}} = 0.22 \text{ Hz}. \tag{4.36}$$

Let us evaluate the components of uncertainty contributed by technical limitations and by the deviation of the ambient temperature from the normal conditions.

The component of the instrumental uncertainty of Type B, introduced by the frequency MI due to its limited accuracy γ, is estimated using the equation

$$u_{B1} = \frac{\gamma f_{\max}}{100\%\sqrt{3}} = 0.1\frac{51.9}{173} = 0.075 \text{ Hz}. \tag{4.37}$$

The component of uncertainty caused by the frequency variation of the quartz resonator due to the deviation of the ambient temperature from the normal value $T_n = 20°\text{C}$ to $T_m = 23°\text{C}$ is estimated using the temperature coefficient of the quartz resonator frequency $\left(k_T = \pm 1.5 \cdot 10^{-6}/°\text{C}\right)$ according to the following equation:

$$u_{B2} = \bar{f}\frac{|T_m - T_n|}{\sqrt{3}}k_T = 50.1\frac{|23-20|}{\sqrt{3}}1.5 \cdot 10^{-6} = 0.13 \cdot 10^{-3} \text{ Hz}. \tag{4.38}$$

The combined Type B standard uncertainty of the measurement result is estimated using the following equation:

$$u_{cB} = \sqrt{u_{B1}^2 + u_{B2}^2} = \sqrt{\left(75 \cdot 10^{-3}\right)^2 + \left(0.13 \cdot 10^{-3}\right)^2} = 0.075 \text{ Hz}. \tag{4.39}$$

As the calculations show, the component of the combined Type B uncertainty u_{B2} is negligible and can be disregarded.

The combined standard uncertainty of the result of repeated measurements of the sinusoidal signal frequency of a general-purpose electrical network is calculated using the following

$$u_c = \sqrt{u_A^2(\bar{f}) + u_{cB}^2} = 0.23 \text{ Hz}. \tag{4.40}$$

Given a confidence level of $p = 0.95$ (coverage factor $k_p = 1.96$), the expanded uncertainty is calculated using the following equation:

$$U = k_p u_c = 1.96 \cdot 0.23 = 0.45 \text{ Hz}. \tag{4.41}$$

The result of the repeated measurement of the sinusoidal signal frequency of a general-purpose electrical network is presented as follows:

$$f = \bar{f} \pm U = 50.1 \pm 0.45 \text{ Hz, for } p = 0.95. \tag{4.42}$$

4.4.3 Evaluation of the Measurement Uncertainty of the Density of a Solid

It is required to evaluate the uncertainty of the density ρ of a solid based on the results of measuring its volume V and mass M at a confidence level of $p = 0.95$.

The mass of the body was measured by repeated weighing using a set of standard weights, with an error not exceeding $\Delta_M = 0.01$ mg. The volume of the body was determined using the hydrostatic weighing method with the same set of weights. The results of the volume and mass measurements are presented in Table 4.4.

The measurement equation is expressed as:

$$\rho = \frac{M}{V}. \tag{4.43}$$

Let us determine the mean values of the observed results for volume and mass:

$$\bar{V} = \frac{\sum_{i=1}^{n} V_i}{n} = 195.38 \cdot 10^{-6} \text{ m}^3. \tag{4.44}$$

$$\bar{M} = \frac{\sum_{i=1}^{n} M_i}{n} = 252.91 \cdot 10^{-3} \text{ kg}. \tag{4.45}$$

TABLE 4.4
Results of Volume and Mass Measurements

i	Volume of the Body $V_i \cdot 10^{-6}$, m^3	Mass of the Body $M_i \cdot 10^{-3}$, kg
1	195.3799	252.9119
2	195.3830	252.9133
3	195.3790	252.9151
4	195.3819	252.9130
5	195.3795	252.9109
6	195.3788	252.9094
7	195.3792	252.9113
8	195.3794	252.9115
9	195.3791	252.9119
10	195.3791	252.9115
11	195.3794	252.9118

The measurement result for the density of the body is determined using Equation (4.43):

$$\overline{\rho} = \frac{\bar{M}}{\overline{V}} = 1.29 \cdot 10^3 \text{ kg/m}^3.$$

Using Type A evaluation, the standard uncertainties of the volume and mass measurements are determined using the following expressions:

$$u_A(\bar{V}) = \sqrt{\frac{\sum_{i=1}^{n}(V_i - \bar{V})^2}{n(n-1)}} = 4.05 \cdot 10^{-10} \text{ m}^3; \tag{4.46}$$

$$u_A(\bar{M}) = \sqrt{\frac{\sum_{i=1}^{n}(M_i - \bar{M})^2}{n(n-1)}} = 4.4 \cdot 10^{-7} \text{ kg}. \tag{4.47}$$

The combined Type A uncertainty of the indirect density measurement is calculated using the following equation:

$$u_{cA}(\rho) = \sqrt{\left(\frac{\partial\rho}{\partial V}\right)^2 u_A^2(\bar{V}) + \left(\frac{\partial\rho}{\partial M}\right)^2 u_A^2(\bar{M})}, \tag{4.48}$$

where $\frac{\partial\rho}{\partial V} = -\frac{M}{V^2} = -6625.37 \cdot 10^3 \frac{kg}{m^6}$ and $\frac{\partial\rho}{\partial M} = \frac{1}{V} = 5118.24 \frac{1}{m^3}$ are the sensitivity coefficients.

By substituting the values of the Type A standard uncertainties and the sensitivity coefficients into Equation (4.48), we obtain the combined uncertainty equal to $u_{cA}(\rho) = 3.5 \cdot 10^{-3}$ kg/m^3.

The Type B standard uncertainty, caused by the presence of residual systematic errors of the standard weights, assuming a uniform distribution of the error, is calculated as follows:

$$u_B = \frac{\Delta_M}{\sqrt{3}} = \frac{0.01 \cdot 10^{-3}}{1.73} = 5.78 \cdot 10^{-6} \text{ kg}. \quad (4.49)$$

The Type B standard uncertainty of the density measurement, taking into account the sensitivity coefficient with respect to mass $\frac{\partial \rho}{\partial M} = \frac{1}{V} = 5118.2364 \frac{1}{m^3}$, is calculated as follows:

$$u_{cB} = \sqrt{\left(\frac{\partial \rho}{\partial M}\right)^2 u_B^2} = \sqrt{(5118.24)^2 \cdot \left(5.78 \cdot 10^{-6}\right)^2} = 29.58 \cdot 10^{-3} \text{ kg/m}^3. \quad (4.50)$$

The combined uncertainty of the solid body density measurement, considering both Type A and Type B standard uncertainties, is determined as follows:

$$u_c = \sqrt{u_{cA}^2(\rho) + u_{cB}^2} =$$

$$= \sqrt{\left(3.5 \cdot 10^{-3}\right)^2 + \left(29.58 \cdot 10^{-3}\right)^2} = 29.79 \cdot 10^{-3} \text{ kg/m}^3. \quad (4.51)$$

The expanded uncertainty of the density estimate is calculated using the following expression:

$$U = k_p u_c, \quad (4.52)$$

where k_p is the coverage factor, determined based on the confidence level $p = 0.95$ and the effective degrees of freedom, which is calculated using the Welch–Satterthwaite equation:

$$v_{eff} = \frac{u_c^4}{\sum_{i=1}^{N} \frac{u_i^4}{v_i}} = \frac{u_c^4}{\frac{u_{cA}^4(\rho)}{n-1} + \frac{u_{cB}^4}{\infty}} = 10 \frac{29.79^4}{3.5042^4} = 52230 \rightarrow \infty. \quad (4.53)$$

Thus, the expanded uncertainty (4.52) is equal to: $U = 1.96 \cdot 29.79 \cdot 10^{-3} = 58.39 \cdot 10^{-3} \text{ kg/m}^3$.

Therefore, the result of the indirect measurement of the solid body density can be expressed as:

$$\rho = \overline{\rho} \pm U = 1.29 \cdot 10^3 \pm 58.39 \cdot 10^{-3} \text{ kg/m}^3, \text{for } p = 0.95. \quad (4.54)$$

4.4.4 Evaluation of Measurement Uncertainty of Gluten Deformation

As a result of the calibration of the gluten deformation gauge, according to the calibration procedure specified in the instrument's manual, the following measurement results were obtained in arbitrary units:

- Using a gauge block of 2 mm thickness, the deformation values are: 121.6; 122.3; 122.5.
- Using a gauge block of 6 mm thickness, the deformation values are: 64.5; 65.1; 65.5.
- Using a gauge block of 10 mm thickness, the deformation values are: 7.5; 7.9; 8.4.

According to the instrument manual, the following information is known:

- The measurement range of gluten deformation is from 0 to 150.7 arbitrary units.
- The primary absolute measurement error does not exceed $\Delta_{pr} = \pm 0.5$ arbitrary units.
- The permissible absolute additional measurement error due to deviation of the mains voltage from the nominal value does not exceed $\Delta_{add} = \pm 0.5$ arbitrary units.
- The permissible absolute measurement error due to deviation of the ambient temperature from nominal values within +10 to +35°C does not exceed $\Delta_T = \pm 0.5$ arbitrary units.

The mean values of gluten deformation obtained from the calibration are:

$$\bar{q}_1 = \frac{\sum_{k=1}^{n} q_k}{n} = \frac{121.6 + 122.3 + 122.5}{3} = 122.13\ a.u.; \tag{4.55}$$

$$\bar{q}_2 = \frac{65.5 + 65.1 + 65.4}{3} = 65\ a.u.;$$

$$\bar{q}_3 = \frac{7.5 + 7.9 + 8.4}{3} = 7.93\ a.u.$$

The Type A standard uncertainty is calculated based on the measurement results obtained during calibration using the equation

$$u_A = \sqrt{\frac{\sum_{k=1}^{n} (q_k - \bar{q})^2}{n(n-1)}}. \tag{4.56}$$

The experimental measurement uncertainties of the dough deformation are as follows:

for the first group $u_{A1} = 0.27\ a.u.$;
for the second group $u_{A2} = 0.29\ a.u.$;
for the third group $u_{A3} = 0.26\ a.u.$

The Type B standard uncertainty, caused by the presence of the main absolute measurement error of deformation Δ_{pr}, assuming a uniform distribution of the error, is calculated as follows

$$u_{B1} = \frac{\Delta_{pr}}{\sqrt{3}} = \frac{0.5}{1.73} = 0.29\ a.u. \tag{4.57}$$

The Type B standard uncertainty, introduced due to the additional measurement error Δ_{add} caused by deviations of the mains voltage from its nominal value, assuming a uniform distribution of the error, is calculated similarly:

$$u_{B2} = \frac{\Delta_{\text{add}}}{\sqrt{3}} = \frac{0.5}{1.73} = 0.29\ a.u. \tag{4.58}$$

The Type B standard uncertainty, caused by deviations of the ambient temperature from the nominal value, assuming a uniform distribution of the error, is given by

$$u_{B3} = \frac{\Delta_T}{\sqrt{3}} = \frac{0.5}{1.73} = 0.29\ a.u. \tag{4.59}$$

The combined Type B uncertainty, arising from sources of uncertainty that have a systematic nature, is calculated as follows:

$$u_{Bc} = \sqrt{u_{\text{B1}}^2 + u_{\text{B2}}^2 + u_{\text{B3}}^2} = 0.5\ a.u. \tag{4.60}$$

The combined experimental Type A uncertainty, arising from sources of uncertainty that have a random nature, is calculated using the following expression

$$u_{Ac} = \sqrt{u_{\text{A1}}^2 + u_{\text{A2}}^2 + u_{\text{A3}}^2} = 0.47\ a.u. \tag{4.61}$$

The combined uncertainty of the gluten deformation measurement result, arising from both sources of uncertainty with a random nature and sources of uncertainty with a systematic nature, is calculated using the following expression

$$u_c = \sqrt{u_{\text{Ac}}^2 + u_{\text{Bc}}^2} = 0.69\ a.u. \tag{4.62}$$

The expanded uncertainty of the gluten deformation measurement is determined using the coverage factor k_p, which is selected from the table corresponding to the confidence level p and the effective number of degrees of freedom ν_{eff}, calculated according to the Welch–Satterthwaite equation:

$$\nu_{\text{eff}} = \frac{u_c^4}{\sum_{i=1}^{N} \frac{u_i^4}{\nu_i}} = \frac{u_c^4}{\frac{u_{A1}^4}{n-1} + \frac{u_{A2}^4}{n-1} + \frac{u_{A3}^4}{n-1} + \frac{u_{Bc}^4}{\infty}} = 14. \tag{4.63}$$

For a confidence level of $p = 0.95$ and an effective number of degrees of freedom 14, the coverage factor is taken from Student's t-table and equals $k_p = 2.145$.

Thus, the expanded uncertainty that can be attributed to the gluten deformation MI based on the calibration results is:

$$U = 2.145 \cdot 0.69 = 1.48\ a.u., \text{for } p = 0.95. \tag{4.64}$$

Considering the value of the expanded uncertainty, the measurement result can be presented as:

$$q_1 = \bar{q}_1 \pm U = 122.13\ a.u. \pm 1.48\ a.u., \text{for } p = 0.95; \tag{4.65}$$

$$q_2 = \bar{q}_2 \pm U = 65\ a.u. \pm 1.48\ a.u., \text{for } p = 0.95; \tag{4.66}$$

$$q_3 = \bar{q}_3 \pm U = 7.93\ a.u. \pm 1.48\ a.u., \text{for } p = 0.95. \tag{4.67}$$

4.4.5 Evaluation of the Measurement Uncertainty of Vibration Acceleration during Accelerometer Calibration

As a result of calibrating a piezoelectric accelerometer at the upper measurement range, it is subjected to a reference vibration acceleration of 1 g (10 m/s^2). The output voltage of the accelerometer, measured using a precision millivoltmeter, yielded the following values in millivolts ($i = 1, \ldots, n; n = 10$):

0.5; 0.52; 0.51; 0.49; 0.5; 0.49; 0.51; 0.52; 0.5; 0.51.

The transfer equation of the piezoelectric accelerometer is given by:

$$g = \frac{U}{MS} = \frac{CU}{2d_{11}M}, \tag{4.68}$$

where U is output voltage of the accelerometer; M is mass of the accelerometer; $S = 2d_{11}/C$ is accelerometer sensitivity; C is capacitance of the piezoelectric transducer; d_{11} is piezoelectric constant describing the primary sensitivity of the piezoelectric charge generated when a force is applied $\left(d_{11} = 2.1 \cdot 10^{-12}\ \text{F} \cdot \text{V/N}\right)$.

According to the manufacturer's specification, the combined error of the accelerometer is $\gamma = \pm 1\%$, the accelerometer mass is $M = 24 \pm 1$ g, the accelerometer sensitivity is $S = 2.1$ mV/N, and the capacitance instability of the cable is $\Delta C = \pm 80$ pF.

Based on the measured voltage values, the arithmetic mean voltage is calculated using the equation

$$\bar{U} = \frac{\sum_{i=1}^{n} U_i}{n} = 0.51 \text{ mV}. \tag{4.69}$$

The estimate of the vibration acceleration measurement result g is calculated using Equation (4.68):

$$\bar{g} = \frac{\bar{U}}{MS} = \frac{0.51}{0.024 \cdot 2.1} = 10.12 \text{ m/s}^2.$$

The standard uncertainty of the voltage measurement is estimated using the equation

$$u_A(\bar{U}) = \sqrt{\frac{\sum_{i=1}^{n} (U_i - \bar{U})^2}{n(n-1)}} = 3.42 \text{ µV}. \tag{4.70}$$

The standard uncertainty of the vibration acceleration measurement, arising from sources of uncertainty with a random nature, is estimated through the standard uncertainty of the voltage measurement using the equation

$$u_A(\bar{g}) = \frac{\partial g}{\partial U} u_A(\bar{U}) = \frac{C}{2d_{11}M} u_A(\bar{U}) = 67.86 \cdot 10^{-3} \text{ m/s}^2. \tag{4.71}$$

The Type B standard uncertainty, arising from the limited accuracy of the accelerometer ($\gamma = \pm 1\%$), assuming a uniform distribution of the error within the limits, is determined using the equation

$$u_{B,\gamma} = \frac{\gamma}{\sqrt{3}100\%} \bar{U} = \frac{1}{173} 0.51 = 2.95 \text{ µV}. \tag{4.72}$$

The Type B standard uncertainty, caused by the presence of unremoved limits of the systematic component of the accelerometer mass measurement error $\Theta_M = \pm 10^{-3}$ kg, assuming a uniform distribution of the error within the limits, is calculated using the equation

$$u_{B,M} = \frac{\Theta_M}{\sqrt{3}} = \frac{10^{-3}}{1.73} = 0.58 \ g. \tag{4.73}$$

The Type B standard uncertainty, caused by the instability of the cable capacitance ΔC, is calculated using the equation

$$u_{B,C} = \frac{\Delta C}{\sqrt{3}} = \frac{80}{1.73} = 46.24 \ pF. \tag{4.74}$$

The combined Type B uncertainty, considering the sensitivity coefficients, is calculated using the equation

$$u_{c_B} = \sqrt{\left(\frac{\partial g}{\partial U}\right)^2 u_{B,\gamma}{}^2 + \left(\frac{\partial g}{\partial M}\right)^2 u_{B,M}{}^2 + \left(\frac{\partial g}{\partial C}\right)^2 u_{B,C}{}^2} =$$

$$= \sqrt{\left(\frac{C}{2d_{11}M}\right)^2 u_{B,\gamma}{}^2 + \left(-\frac{CU}{2d_{11}M^2}\right)^2 u_{B,M}{}^2 + \left(\frac{U}{2Md_{11}}\right)^2 u_{B,C}{}^2}. \tag{4.75}$$

Substituting the values of the sensitivity coefficients and the corresponding Type B standard uncertainties into Equation (4.75), we obtain the combined Type B uncertainty equal to $u_{c_B} = 0.36 \cdot 10^{-12}\ \mathrm{m/s^2}$.

The combined uncertainty of the vibration acceleration measurement result is calculated using the equation

$$u_c = \sqrt{u_A^2(\bar{g}) + u_{c_B}^2} = 0.07\ \mathrm{m/s^2}. \tag{4.76}$$

The coverage factor is determined based on the effective number of degrees of freedom and the confidence level $p = 0.95$

$$\nu_{eff} = (n-1)\frac{u_c^4}{u_A^4} = 9\frac{0.07^4}{67.86^4 \cdot 10^{-12}} = 10. \tag{4.77}$$

For $p = 0.95$ and $\nu_{\mathrm{eff}} = 10$, the coverage factor is $k_p = 2.228$.

Then, the expanded uncertainty of the indirect measurement of vibration acceleration during accelerometer calibration is calculated using the equation

$$U_{0.95} = k_p u_c = 0.16\ \mathrm{m/s^2}. \tag{4.78}$$

Hence, the result of the vibration acceleration measurement can be expressed as

$$g = 10.12 \pm 0.16\ \mathrm{m/s^2}, \text{for } p = 0.95. \tag{4.79}$$

4.4.6 Evaluation of the Measurement Uncertainty of Force

The force measurement was carried out using a transducer, which is described by the following conversion equation:

$$Q = 4R\frac{hS}{U_g U_b}\Delta UX, \tag{4.80}$$

where Q is the force (measurand), ΔU is the output voltage of the force sensor (input voltage of the scaling converter); U_g is the output voltage of the scaling converter; $k = U_g/\Delta U = 40$ is the gain of the operational amplifier, S is the sensitivity of the force sensor, h is the ADC quantization step, U_b is the voltage supplied to the bridge circuit of the force sensor ($U_b = 6.5$ V $\pm$ 0.25 mV), R is the resistance of the bridge circuit of the strain gauges ($R = 3.25 \pm 0.03$ kΩ), and X is the number of pulses at the ADC output.

Measurements were carried out under laboratory conditions at an ambient temperature of + 25°C.

According to the technical documentation of the measurement device:

- the zero-offset voltage drift of the scaling converter (operational amplifier) is $U_{\text{drift}} = 3\ \mu\text{V}/^{\circ}\text{C}$;
- ADC resolution $m = 10$ bits;
- permissible deviation of the force sensor supply voltage is $\Theta_{Ub} = 0.25$ mV;
- permissible deviation of the resistance of the force sensor bridge is $\Theta_R = 30$ Ω;
- temperature coefficient of resistance is $\alpha_t = 2.5 \cdot 10^{-6}\ \text{K}^{-1}$.

It is required to evaluate the measurement uncertainty of the force based on the observed values Q_i in kilograms ($I = 1, \ldots, n; n = 16$):

10.1; 10.3; 10.1; 10.2; 9.8; 10; 10.2; 10.5; 10.1; 9.7; 9.8; 9.9; 10.3; 10.4; 10; 9.7.

The evaluation of the measured force result is calculated using the equation:

$$\overline{Q} = \frac{\sum_{i=1}^{n} Q_i}{n} = 10.07\ \text{kg}. \tag{4.81}$$

The Type A standard uncertainty of the measured force results Q_i, which arises from uncertainty sources of a random nature, is evaluated using the following equation:

$$u_A(Q_i) = \sqrt{\frac{\sum_{i=1}^{n} \left(Q_i - \overline{Q}\right)^2}{n-1}} = 0.24\ \text{kg}. \tag{4.82}$$

Let us check the measurement results for the presence of outliers. For this purpose, we calculate the ratio:

$$s_1 = \frac{\bar{Q} - Q_{\min}}{u_A(Q_i)} = 1.54; \tag{4.83}$$

$$s_2 = \frac{Q_{\max} - \bar{Q}}{u_A(Q_i)} = 1.79. \tag{4.84}$$

We adopt a confidence level of $p = 0.95$ and determine the significance level as $\alpha = 1 - p = 1 - 0.95 = 0.05$. Depending on the number of performed measurements ($n = 16$), the permissible value of the normalized deviation is selected from Table 3.1, which in this case is $s_{crit} = 2.44$.

Since the values of the ratios s_1 and s_2, calculated according to Equations (4.83) and (4.84), are less than the permissible tabulated value s_{crit}, the hypothesis of the presence of abnormal results in the set of observations is rejected. If the value of one or both ratios had exceeded the permissible value s_{crit}, it would have been necessary to remove the minimum or maximum observation, as this would indicate the presence of outliers in the measurement results.

The Type A standard uncertainty of the mean value is calculated using the following equation:

$$u_A(\overline{Q}) = \sqrt{\frac{\sum_{i=1}^{n} (Q_i - \overline{Q})^2}{n(n-1)}} = 0.06\ \text{kg}. \tag{4.85}$$

Next, we proceed to the evaluation of Type B standard uncertainties, which arise from the technical limitations of the force MI.

Since the force measurements were performed under laboratory conditions at an ambient temperature of $T = +25°\text{C}$, which differs from the standard reference temperature $T_{ref} = 20°\text{C}$, and the specification of the force measuring device indicates that the zero-offset voltage drift of the operational amplifier is $U_{drift} = 3\ \mu\text{V}/°\text{C}$, the Type B standard uncertainty due to the zero-offset voltage drift of the operational amplifier, assuming a uniform distribution of the error within the limits, is determined using the following equation:

$$u_{B,T} = k\frac{|T - T_{ref}|}{\sqrt{3}} U_{drift} = 40\frac{|25 - 20|}{\sqrt{3}} 3 = 0.35\ \text{mV}. \tag{4.86}$$

The Type B standard uncertainty, caused by the limited resolution of the ADC, assuming a triangular distribution of the error within the limits, is determined using the following equation:

$$u_{B,h} = \frac{h}{\sqrt{24}} = \frac{U_{ref}}{(2^m - 1)\sqrt{24}} = 0.6\ \text{mV}, \tag{4.87}$$

where U_{ref} is the reference voltage of the ADC ($U_{ref} = 3$ V).

The Type B standard uncertainty, caused by the deviation of the force sensor supply voltage from its nominal value, is determined using the following equation:

$$u_{B,\theta_{Ub}} = \frac{\theta_{Ub}}{\sqrt{3}} = \frac{0.25}{1.73} = 0.14\ \text{mV}. \tag{4.88}$$

The Type B standard uncertainty, caused by the residual systematic error of the bridge circuit resistance, is calculated using the following equation:

$$u_{B,\theta_R} = \frac{\theta_R}{\sqrt{3}} = \frac{30}{1.73} = 17.34\ \Omega. \tag{4.89}$$

The Type B standard uncertainty, caused by the temperature error of the strain gauges when the ambient temperature deviates from standard conditions, can be determined based on the equation describing the resistance–temperature dependence $R = R_0 \cdot \alpha_t \cdot \Delta t$, using the following equation:

$$u_{B,Rt} = \alpha_t \frac{|\Delta t|}{\sqrt{3}} R = 2.5 \cdot 10^{-6} \frac{|25 - 20|}{\sqrt{3}} 3.25 \cdot 10^3 = 0.02\ \Omega. \tag{4.90}$$

The combined Type B uncertainty is calculated using the following equation:

$$u_{Bc} = \left[\left(\frac{\partial Q}{\partial U_g} \right)^2 u_{B,t}{}^2 + \left(\frac{\partial Q}{\partial h} \right)^2 u_{B,h}{}^2 + \left(\frac{\partial Q}{\partial U_b} \right)^2 u_{B,\theta_{Ub}}{}^2 + \left(\frac{\partial Q}{\partial R} \right)^2 \left(u_{B,\theta_R}{}^2 + u_{B,R}{}^2 \right) \right]^{0.5}, \tag{4.91}$$

where $\frac{\partial Q}{\partial U_g} = -\frac{4RhS}{kU_g U_b} X = -0.2\ \mathrm{kg/V}$, $\frac{\partial Q}{\partial h} = \frac{4RS}{kU_b} X = 206.8\ \mathrm{kg/V}$, $\frac{\partial Q}{\partial U_b} = -\frac{4RhS}{kU_b{}^2} X = -0.16\ \mathrm{kg/V}$, and $\frac{\partial Q}{\partial R} = \frac{4hS}{kU_b} X = 0.3 \cdot 10^{-3}\ \mathrm{kg/\Omega}$ are sensitivity coefficients.

By substituting the calculated values of the sensitivity coefficients and the corresponding Type B standard uncertainties into Equation (4.91), we obtain the value of the combined Type B uncertainty, which equals $u_{Bc} = 0.16$ kg.

The combined uncertainty of the force measurement result is calculated based on the experimental Type A standard uncertainty and the combined Type B uncertainty using the following equation:

$$u_c = \sqrt{u_A^2(\overline{Q}) + u_{Bc}^2} = 0.17\ \mathrm{kg}. \tag{4.92}$$

The effective number of degrees of freedom is calculated using equation:

$$\nu_{\mathrm{eff}} = \frac{u_c^4}{\sum_{i=1}^{N} \frac{u_i^4}{\nu_i}} = (n-1) \frac{u_c^4}{u_A^4(\overline{Q})} = 15 \frac{0.17^4}{0.06^4} = 967. \tag{4.93}$$

The coverage factor k_p is determined based on the effective number of degrees of freedom and the confidence probability, which is taken as $p = 0.95$, and equals $k_p = 1.96$.

The expanded uncertainty of the force measurement result is calculated based on the values of the combined measurement uncertainty and the coverage factor using the following equation:

$$U_{0.95} = k_p u_c = 0.33\ \mathrm{kg}. \tag{4.94}$$

TABLE 4.5
Uncertainty Budget

Input Quantity	Estimate of Input Quantity	Standard Uncertainties	Sensitivity Coefficients	Uncertainty Contributions
Q_i	$\overline{Q}$	0.06 kg	1	$u_A(\overline{Q})$
U_{drift}	3 μV/°C	0.35 mV	−0.2 kg/V	$\frac{\partial Q}{\partial U_g} u_{B,t}$
h	2.93 mV	0.6 mV	206.8 kg/V	$\frac{\partial Q}{\partial h} u_{B,h}$
θ_{Ub}	0.25 mV	0.14 mV	−0.16 kg/V	$\frac{\partial Q}{\partial U_b} u_{B,\theta_{Ub}}$
θ_R	30 Ω	17.34 Ω	$0.3 \cdot 10^{-3}$ kg/Ω	$\frac{\partial Q}{\partial R} u_{B,\theta_R}$
α_t	$2.5 \cdot 10^{-6}$ K^{-1}	0.02 Ω	$0.3 \cdot 10^{-3}$ kg/Ω	$\frac{\partial Q}{\partial R} u_{B,Rt}$
Output Quantity	**Estimate of Output Quantity**	**Combined Uncertainty**	**Coverage Factor**	**Expanded Uncertainty**
$\overline{Q}$	10.07 kg	0.17 kg	1.96	0.33 kg

The force measurement result is presented in the form:

$$Q = \overline{Q} \pm U_{0.95} = 10.07 \pm 0.33 \text{ kg, for } p = 0.95. \tag{4.95}$$

Based on the calculated uncertainties, an uncertainty budget can be compiled, as shown in Table 4.5.

4.4.7 Evaluation of the Measurement Uncertainty of the Gauge Pressure of a Gaseous Medium Using the Manometer

The measurement of the gauge pressure of a gaseous medium is carried out with repeated observations using the manometer.

The manometer readings P, in MPa, are as follows:

15; 16; 15; 14; 15; 16; 17; 14; 15; 16; 14; 15; 16; 15; 13; 16; 17; 16; 15; 16.

It is required to evaluate the measurement uncertainty of pressure at a confidence level $p = 0.95$, under the assumption of a normal distribution of the measurement results.

We compile the measurement specification:

1. Analysis of the measurement conditions:
   a. measurements were performed under laboratory conditions at an ambient temperature of +20°C;

2. Analysis of the technical characteristics of the MI:
   a. operating conditions of the MI: ambient temperature from –30 to +50°C;
   b. upper measurement limit: $P_{max} = 25$ MPa;
   c. accuracy class of the pressure MI: 2.5.

Let us check the measurement results for the presence of gross outliers. For this purpose, we calculate:

- the arithmetic mean of the obtained results:

$$\bar{P} = \frac{\sum_{k=1}^{n} P_k}{n} = \frac{\sum_{k=1}^{20} P_k}{20} = 15.3 \text{ MPa}. \tag{4.96}$$

- the SD of the results from the arithmetic mean value (Type A uncertainty):

$$u_A = \sqrt{\frac{\sum_{k=1}^{n} (P_k - \bar{P})^2}{n-1}} = 1.03 \text{ MPa}. \tag{4.97}$$

- the interval of the expanded uncertainty corresponding to the confidence level of 0.95, under the assumption of a normal distribution of the observation results:

$$U = 1.96u_A = 2.02 \text{ MPa}. \tag{4.98}$$

- the limits of this interval for the observation results are:

$$P_{min} = 13.28 \text{ MPa}; P_{max} = 17.32 \text{ MPa}.$$

Thus, as the calculations show, the smallest observation with a value of 13 MPa falls outside the confidence interval and is therefore removed from the set of observations as being affected by a blunder or gross error.

Accordingly, after checking the measurement results for outliers and removing them from the observations, we compute the arithmetic mean of the corrected observation results, which is taken as the measurement result:

$$\bar{P} = \frac{\sum_{k=1}^{19} P_k}{19} = 15.42 \text{ MPa}.$$

Next, we calculate the experimental SD of the observation result (Type A uncertainty), which equals:

$$u_A = \sqrt{\frac{\sum_{k=1}^{19} (P_k - \bar{P})^2}{19-1}} = 0.9 \text{ MPa}.$$

By checking again for outliers using Equation (4.98), we confirmed that there are no further outliers in the observation results.

Since the mean value is taken as the result of repeated measurements, it is important to evaluate its variance. The best estimate of the variance of the mean value is expressed by Equation (1.3). Therefore, the experimental Type A uncertainty of the measurement result (the mean value) is:

$$u_A(\bar{P}) = \sqrt{\frac{u_A^2}{n}} = 0.21 \text{ MPa}. \tag{4.99}$$

The relative Type A standard uncertainty is calculated using Equation (1.51), and we obtain:

$$\tilde{u}_A = \frac{u_A(\bar{P})}{\bar{P}} 100\% = \frac{0.21}{15.42} 100\% = 1.36\%. \tag{4.100}$$

Next, we proceed to the evaluation of the Type B standard uncertainty, which arises from the technical limitations of the MI.

As is known, Type B uncertainty may be evaluated based on information provided in the manufacturer's specification (datasheet), reference data, data from previous measurements, information on the form of the probability distribution, etc.

Since the manufacturer's specification states that the accuracy class of the pressure MI is $\gamma = 2.5$, the Type B standard uncertainty, under the assumption of a uniform distribution of the error introduced by the MI due to its limited accuracy γ, can be evaluated using the following equation:

$$u_B = \frac{|\Delta|}{\sqrt{3}}, \tag{4.101}$$

where Δ is the limits of the absolute error, which in this case can be calculated from the accuracy class using the following equation:

$$\Delta = \frac{\gamma P_H}{100\%} = \frac{2.5}{100} 25 = 0.63 \text{ MPa}. \tag{4.102}$$

By substituting the calculated value of the absolute error into the equation for determining the Type B standard uncertainty (4.101) introduced by the pressure MI due to the limited accuracy class γ, we obtain a Type B standard uncertainty equal to 0.36 MPa, and the relative Type B standard uncertainty equals $\tilde{u}_B = u_B 100\%/\bar{P} = 2.33\%$.

The combined standard uncertainty of the pressure measurement result is calculated using the following equation:

$$u_c = \sqrt{u_A^2(\bar{P}) + u_B^2} = \sqrt{0.21^2 + 0.36^2} = 0.42 \text{ MPa}. \tag{4.103}$$

The effective number of degrees of freedom is determined for the calculation of the expanded uncertainty of the measurement result using the following equation:

$$\nu_{\text{eff}} = \frac{u_c^4}{\frac{u_A^4}{\nu_i} + \frac{u_B^4}{\infty}} = (n-1)\frac{u_c^4}{u_A^4} = (19-1)\frac{0.42^4}{0.21^4} = 288. \tag{4.104}$$

Assuming a confidence level of $p = 0.95$, we calculate the expanded uncertainty using Equation (1.47), considering the number of measurements $n = 19$ and the effective number of degrees of freedom, which is $\nu_{\text{eff}} \gg 30$. Student's t-distribution factor is then taken as 1.96.

Thus, the expanded uncertainty of the pressure measurement result is:

$$U_{0.95} = 1.96u_c = 0.82 \text{ MPa}. \tag{4.105}$$

Therefore, the pressure measurement result is expressed in the form:

$$P = \bar{P} \pm U_p = 15.42 \pm 0.82 \text{ MPa, at } p = 0.95 \tag{4.106}$$

or

$$14.6 \leq P \leq 16.24 \text{ MPa, at } p = 0.95.$$

4.4.8 Evaluation of the Measurement Uncertainty of Ion Activity

The model equation of the measurement channel (MC) for ion activity with an ion-selective transducer is given by:

$$\Delta U = U_0 + \frac{2.3RT}{n_A F} lg\left(a_A + K_c(a_B)^{n_A/n_B}\right), \tag{4.107}$$

where ΔU is the potential difference at the output of the measuring transducer, U_0 is the standard constant potential of the sensitive element (reference electrode) ($U_0 = 201 \pm 3$ mV), R is the universal gas constant, T is the absolute temperature, F is the Faraday constant, n_A, n_B are charges of ions A and B, respectively, a_A is the activity of ions A, which is to be determined, a_B is e of ions B, which interferes with the determination of ions A, and K_c is the selectivity coefficient (its maximum possible value is 10^{-1}).[6]

Since in monitoring the humus state of soils it is necessary to measure the content of such monovalent substances as fluoride, nitrate nitrogen, ammonium nitrogen, and potassium, in Equation (4.107) the ratio $\frac{2.3RT}{n_A F}$ is a constant that reflects the sensitivity of the ion-selective electrodes to ions A. This sensitivity is $S = 59.16$ mV at the temperature of the studied medium $T = 25°$C.

Changes in ion activity lead to changes in the potential difference. For further amplification of small potential differences in the ion activity MC, a measuring amplifier is used, which must have:

- a differential input to reduce the effect of common-mode interference;
- a low zero-level signal;
- a high common-mode rejection ratio.

After amplification, the measurement signal is fed to the input of an analog-to-digital converter (ADC), where it is converted into a binary code N.

Thus, the final form of the ion activity MC conversion equation becomes:

$$N = \left[U_0 + Slg(a_A + 0.1a_B)\right]\frac{K2^m}{U_{\text{ref}}}, \tag{4.108}$$

where U_{ref} is the reference voltage of the ADC (5 V), m is the resolution of the ADC ($m = 16$), and K is the amplification factor.

From previous studies, it is known that the limits of residual residual systematic errors are as follows:

- the standard potential U_0 of the sensitive element is $\theta_{U0} = \pm 3$ mV;
- the absolute error of the ion activity measurement is $\theta_a = 5 \cdot 10^{-6} \frac{\text{mol}}{\text{dm}^3}$;
- the maximum temperature drift coefficient of the amplifier is 0.25 mV/°C;
- the temperature coefficient of the ADC reference voltage is $k_t = 10^{-5}/^{\circ}C$;
- the deviation of the reference voltage from its nominal value does not exceed $\theta_{U_{\text{ref}}} = \pm 0.02$ V at $T = 20 \pm 5$°C.

Since the distribution law of the uncertainty of the sensitive element is unknown from reference data, it is assumed to be uniformly distributed, and it is calculated using the following equation:

$$u_{U0} = \frac{\theta_{U0}}{\sqrt{3}} = 1.73 \text{ mV}. \tag{4.109}$$

The standard uncertainty arising from the presence of the absolute error in the measurement of ion activity, assuming a normal distribution of the measurement error within the limits, is calculated using the following equation:

$$u_a = \frac{\theta_a}{k_p} = 2.55 \cdot 10^{-3} \text{ mmol/dm}^3, \tag{4.110}$$

where k_p is the coverage factor for a normal distribution, which equals 1.96 for a 0.95 probability.

The combined uncertainty of the ion-selective transducer, arising from the residual residual systematic effects associated with the deviation of the standard potential and the presence of absolute measurement error, calculated as Type B and considering Equations (4.109) and (4.110), is determined using equation:

$$u_{ctr}^2 = \left(\frac{\partial \Delta U}{\partial U_0}\right)^2 u_{U0}^2 + \left(\frac{\partial \Delta U}{\partial a_A}\right)^2 u_a^2 + \left(\frac{\partial \Delta U}{\partial a_B}\right)^2 u_a^2, \tag{4.111}$$

where $\frac{\partial \Delta U}{\partial U_0} = 1$ is the sensitivity coefficient of the standard potential, $\frac{\partial \Delta U}{\partial a_A} = \frac{S}{2.3(a_A + K_c a_B)}$ is sensitivity coefficient of ion A activity, which for the lower measurement range $\left(R_{\min} = 10^{-6}\ \mathrm{mol/dm^3}\right)$ equals 23383.4 V·dm³/mol, $\frac{\partial \Delta U}{\partial a_B} = \frac{K_c S}{2.3(a_A + K_c a_B)}$ is the sensitivity coefficient of interfering ion B activity, which for the lower measurement range $\left(R_{\min} = 10^{-6}\ \mathrm{mol/dm^3}\right)$ equals 2338.34 V·dm³/mol.[7]

By substituting the calculated sensitivity coefficients into Equation (4.111), we obtain the combined uncertainty of the ion-selective transducer, which equals $u_c = 60$ mV.

Since measurements may be performed with ambient temperature varying from 5 to 40°C, the maximum temperature change is $\Delta T = 35$°C. Considering the amplifier temperature drift coefficient of 0.25 mV/°C, the offset voltage is $U_{\text{off}} = 35°\mathrm{C} \cdot 0.25\ \mathrm{mV/^{\circ}C} = 8.75$ mV. With a known amplification factor $K = 10$, the maximum offset voltage will be $U_{\text{off, max}} = KU_{\text{off}} \cdot = 87.5$ mV. Considering the maximum output voltage of the amplifier $U_{\text{out}} = \pm 5$ V, its maximum relative error can be calculated using the following equation:

$$\delta_K = \frac{U_{\text{off, max}}}{2U_{\text{out}}} = 8.75 \cdot 10^{-3}. \tag{4.112}$$

After calculating the maximum relative error of the amplifier (4.112), the standard uncertainty introduced by the amplifier when measuring ion activity, assuming a normal distribution of the error, is calculated using the following equation:

$$u_{B_K} = \frac{\Delta U_{max}}{k_p} \delta_K = 0.83\ \mathrm{mV}, \tag{4.113}$$

where $\Delta U_{\max}$ is the maximum potential difference corresponding to the maximum ion activity (the maximum measurement range of the MC is $R_{\max} = 0.5\ \mathrm{mol/dm^3}$) at the maximum possible activity of interfering ions a_B.

The uncertainty arising from the deviation of the reference voltage from its nominal value, assuming a uniform distribution, is calculated as:

$$u_{U_{\text{ref}}} = \frac{\theta_{U_{\text{ref}}}}{\sqrt{3}} = 11.55\ \mathrm{mV}. \tag{4.114}$$

The additional uncertainty $u_{\Delta T}$, caused by changes in the reference voltage of the source when the ambient temperature varies from $T_1 = 5$°C to $T_2 = 40$°C, which significantly exceeds the normalized temperature deviation, is calculated using the temperature coefficient, assuming a uniform distribution within the limits, with the following equation:

$$u_{\Delta T} = \frac{k_t(T_2 - T_1)}{\sqrt{3}} U_{\text{ref}} = 1.02\ \mathrm{mV}. \tag{4.115}$$

Thus, the combined uncertainty of the ADC reference voltage source, consisting of Type B standard uncertainties calculated using Equations (4.114) and (4.115), is equal to:

$$u_{\text{cref}} = \sqrt{(u_{\text{ref}})^2 + (u_{\Delta T})^2} = 11.59 \text{ mV}. \tag{4.116}$$

The standard uncertainty u_h, caused by the limited quantization step of the ADC voltage, assuming a triangular distribution of the error within the limits, is determined as follows:

$$u_h = \frac{h}{\sqrt{24}} = \frac{U_{ref}}{(2^m - 1)\sqrt{24}} = 15.57 \text{ μV}. \tag{4.117}$$

Considering the Type B uncertainties calculated using Equations (4.111), (4.113), (4.116), and (4.117) at each stage of the measurement signal conversion, the combined uncertainty of the ion activity MC is calculated using the following equation:

$$u_c = \sqrt{u_{ctr}^2 + u_{B_K}^2 + u_{\text{cref}}^2 + u_h^2} = 61.12 \text{ mV}. \tag{4.118}$$

Therefore, analyzing the obtained results, it is evident that the maximum uncertainty arises from the primary ion-selective transducer.

Assuming that the overall distribution of the MC is normal with a confidence level of at least $p = 0.95$, the expanded uncertainty of the ion activity MC is calculated using the following equation:

$$U_{0.95} = k_p u_c = 1.96 \cdot 61.12 = \pm 119.8 \text{ mV}. \tag{4.119}$$

Therefore, based on reference data and previous studies, the expanded uncertainty of the ion activity MC was evaluated, which, in the measurement range from 10^{-6} mol/dm^3 to 0.5 mol/dm^3, does not exceed ±120 mV at the maximum activity of interfering ions a_B.

4.4.9 Evaluation of the Uncertainty of the Measurement Results of Monitoring the Difference of the Rotation Frequencies of Electric Motor Rotors

The measurement equation for the rotation frequencies, followed by the calculation of the difference between the rotation frequencies of the coupled electric motors (EMs; the main and the auxiliary) at given frequencies, is as follows:

$$\Delta N_s = f(A_1, A_2, f_{\text{ref}}, z_m) = N_{s1} - N_{s2} = \frac{60 f_{\text{ref}}}{A_1 z_m} - \frac{60 f_{\text{ref}}}{A_2 z_m}, \tag{4.120}$$

where ΔN_s is the difference of the rotation frequencies of the EM rotors, N_{s1} is the rotation frequency of the rotor of the main EM, N_{s2} is the rotation frequency of the rotor of the auxiliary EM, f_{ref} is the reference value of the sampling frequency $(f_{\text{ref}} = 16\,\text{MHz}), z_m$ is the number of slots of the modulator $(z_m = 30)$, and A_1 and A_2 are the number of pulses corresponding to the rotation frequencies of the main and auxiliary EM.

During the measurement of the number of pulses in accordance with Equation (4.120), three groups of observations of the rotation frequency in revolutions per minute (rpm) were obtained, each group containing 16 values, as presented in Table 4.6.

The measurements were carried out under laboratory conditions at an ambient temperature of +23°C. The operating conditions for the frequency meter are from –30 to +50°C. The relative quantization error δN_s in measuring the rotation frequencies is calculated using the following equation:

$$\delta N_s = \frac{N_s z_m}{60 f_{\text{ref}}} \tag{4.121}$$

TABLE 4.6
Measurement Results of the Main and Auxiliary Motor Rotational Frequencies

	First Group of Observations of the Main and Auxiliary EM Rotational Frequencies		Second Group of Observations of the Main and Auxiliary EM Rotational Frequencies		Third Group of Observations of the Main and Auxiliary EM Rotational Frequencies	
i	N_{s11}	N_{s21}	N_{s12}	N_{s22}	N_{s13}	N_{s23}
1	1195	1183	3005	2993	4889	4866
2	1201	1190	3011	3000	4883	4891
3	1199	1179	3008	2992	4888	4875
4	1188	1207	2999	2971	4869	4883
5	1192	1209	2987	2997	4867	4892
6	1197	1186	2995	3008	4877	4899
7	1199	1178	2998	3010	4850	4871
8	1178	1205	3006	2999	4862	4887
9	1186	1197	2993	2974	4873	4899
10	1194	1171	2996	2988	4859	4878
11	1193	1210	2989	2962	4884	4896
12	1204	1182	3000	2981	4876	4893
13	1200	1189	3012	2999	4855	4882
14	1203	1191	3005	2995	4859	4883
15	1205	1197	3007	3017	4861	4879
16	1198	1196	2994	2973	4877	4894

The relative error of the reference frequency of the internal oscillator is $\delta_{\text{ref}} = \pm 2.5 \cdot 10^{-7}$. The temperature coefficient of the reference oscillator frequency does not exceed $\pm 10^{-9}$ per 1°C.

We will evaluate the components of the combined Type B standard uncertainty associated with the unremoved residual systematic errors of the measurement results.

Since the limits of the relative error of the reference frequency do not exceed $\pm 2.5 \cdot 10^{-7}$, the limits of the absolute error of the measurement of the rotational frequencies will be equal to

$$\Delta_{\text{ref}} = \bar{N}_{s_{\max}} \delta_{\text{ref}} = \pm 2.5 \cdot 10^{-7} \cdot 4878.03 = \pm 1.22 \cdot 10^{-3} \text{ rpm}, \tag{4.122}$$

where $\bar{N}_{s_{\max}} = \frac{\frac{1}{n}\sum_{i=1}^{n} N_{s1i} + \frac{1}{n}\sum_{i=1}^{n} N_{s2i}}{2} = 4878.03$ rpm is the arithmetic mean of the third group of observations at the maximum (upper limit of the measurement range) rotational frequency of the main and auxiliary EM rotors.

The Type B standard uncertainty of the reference frequency u_{B1}, assuming a uniform distribution of the error within the limits, is determined using the following equation:

$$u_{B1} = \frac{\Delta_{ref}}{\sqrt{3}} = 0.71 \cdot 10^{-3} \text{ rpm}. \tag{4.123}$$

The Type B quantization uncertainty u_{B2} is determined from the limits of the relative quantization error (4.121) using the following equation:

$$u_{B2} = \frac{z_m (\bar{N}_{s_{max}})^2}{60 f_{ref} \sqrt{3}} = 0.43 \text{ rpm}. \tag{4.124}$$

The Type B uncertainty u_{B3}, caused by the change of the reference frequency when the ambient temperature varies from +20°C (frequency meter calibration temperature T_c) to +23°C (ambient temperature at the time of measurement T_m), is calculated using the temperature coefficient of frequency $k_t = \pm 1 \cdot 10^{-9}$, assuming a normal distribution, with the following equation:

$$u_{B3} = k_t \frac{|T_m - T_c|}{k_p} \bar{N}_{s_{\max}} = 7.47 \cdot 10^{-6} \text{ rpm}, \tag{4.125}$$

where k_p is the coverage factor for a normal distribution, equal to 1.96 for a confidence level of $p = 0.95$.

The experimental uncertainty of the measurement of the rotational frequencies at the upper measurement limit of the main and auxiliary EMs, caused by sources of uncertainty of a random nature in indirect measurements of the rotational frequencies, considering the correlation between them, is calculated using the following equation:

$$u_A{}^2 = c_1^2 u^2(\bar{N}_{s1_{\max}}) + c_2^2 u^2(\bar{N}_{s2_{\max}}) + \\ + 2 c_1 c_2 u(\bar{N}_{s1_{\max}}) u(\bar{N}_{s2_{\max}}) r(\bar{N}_{s1_{\max}}, \bar{N}_{s2_{\max}}). \tag{4.126}$$

The sensitivity coefficients c_1 and c_2, considering Equation (4.120), are calculated using the following equations:

$$c_1 = \frac{\partial \Delta N_s}{\partial \bar{A}_1} = -\frac{\bar{N}_{s1}{}^2 z_m}{60 f_{\text{ref}}}, c_2 = \frac{\partial \Delta N_s}{\partial \bar{A}_2} = \frac{\bar{N}_{s2}{}^2 z_m}{60 f_{\text{ref}}}, \tag{4.127}$$

The Type A standard uncertainties in the measurement of the rotational frequencies of the main and auxiliary EM at the upper measurement limit are equal to:

$$u_A(\bar{N}_{s1}) = \sqrt{\frac{\sum_{i=1}^{n} (N_{s1\max_i} - \bar{N}_{s1\max})^2}{n(n-1)}}, u_A(\bar{N}_{s2}) = \sqrt{\frac{\sum_{i=1}^{n} (N_{s2\max_i} - \bar{N}_{s2\max})^2}{n(n-1)}}, \tag{4.128}$$

The value of the correlation coefficient between the rotational frequencies of the main and auxiliary EMs is calculated using the following equation:

$$r(\bar{N}_{s1\max}, \bar{N}_{s2\max}) = \frac{\sum_{i=1}^{n} (N_{s1\max_i} - \bar{N}_{s1\max}) \sum_{i=1}^{n} (N_{s2\max_i} - \bar{N}_{s2\max})}{\sqrt{\sum_{i=1}^{n} (N_{s1\max_i} - \bar{N}_{s1\max})^2} \sqrt{\sum_{i=1}^{n} (N_{s2\max_i} - \bar{N}_{s2\max})^2}}. \tag{4.129}$$

To obtain the numerical value of the correlation coefficient, the mean values of the rotational frequencies of the main $\bar{N}_{s1\max} = 4870.56$ rpm and auxiliary $\bar{N}_{s2\max} = 4885.5$ rpm EM at the upper measurement limit (maximum measurement range) were calculated, since in this range the largest deviations of the main motor's rotational frequency from that of the auxiliary motor occur, as well as the largest values of the SDs.

By substituting the arithmetic mean values of the rotational frequencies into Equation (4.127), the following numerical values of the sensitivity coefficients were obtained: $c_1 = -0.74$ rpm and $c_2 = 0.75$ rpm. The standard uncertainty of the rotational frequency measurement of the main EM, considering Equation (4.128), is $u_A(\bar{N}_{s1}) = 3.03$ rpm and for the auxiliary motor—$u_A(\bar{N}_{s2}) = 2.5$ rpm. By substituting these values into Equation (4.129), the correlation coefficient was obtained, which is $r(\bar{N}_{s1\max}, \bar{N}_{s2\max}) = 0.21$.

Thus, substituting the obtained numerical values of the Type A uncertainties and the correlation coefficients into Equation (4.126), the experimental uncertainty of the rotational frequency measurement of the EM rotors is obtained as $u_A = 3.21$ rpm. The relative Type A experimental uncertainty was $\tilde{u}_A = 0.07\%$.

The combined uncertainty of the asynchronism of the EM rotors is calculated using the following equation:

$$u_c = \sqrt{u_{B1}{}^2 + u_{B2}{}^2 + u_{B3}{}^2 + u_A{}^2} = 3.24 \text{ rpm}. \tag{4.130}$$

The expanded uncertainty of the results of the indirect determination of the difference of the rotational frequencies of the rotors is calculated using the following equation:

$$U_{0.95} = t_{0.95}\left(\nu_{eff}\right)u_c. \tag{4.131}$$

The effective number of degrees of freedom ν_{eff} is determined from the following relation:

$$\nu_{\text{eff}} = \frac{u_c{}^4}{\dfrac{u_{B1}{}^4 + u_{B2}{}^4 + u_{B3}{}^4}{\infty} + \dfrac{u_A{}^4}{n-1}} = 16, \tag{4.132}$$

then, the coverage factor is taken from Student's t-table and for an effective number of degrees of freedom of 16, is equal to: $t_{0.95}(\nu_{\text{eff}}) = t_{0.95}(16) = 2.12$.

By substituting the coverage factor and the combined uncertainty into Equation (4.131), the expanded uncertainty of the difference of the rotational frequencies of the main and auxiliary EM rotors is obtained, which is $U_{0.95} = 2.12 \cdot 3.24 = \pm 6.87$ rpm and the relative expanded uncertainty is equal to $\tilde{U}_{0.95} = 0.14\%$.

Therefore, the result of the measurement of the difference of the rotational frequencies of the EM rotors at the maximum rotational frequency of the motors can be expressed as:

$$\Delta N_s = |\bar{N}_{s1_{\max}} - \bar{N}_{s2_{\max}}| = 14.94 \ \text{rpm},$$

$$8.07 \leq \Delta N_s \leq 21.81 \ \text{rpm for} \ p = 0.95,$$

or

$$\Delta N_s \pm U_{0.95} = 14.94 \pm 6.87 \ \text{rpm for} \ p = 0.95.$$

4.4.10 Evaluation of the Uncertainty of the Indirect Measurement of the Rotor Moment of Inertia Using the Amplitude of Torsional Oscillations

The model equation for determining the rotor moment of inertia from the amplitude of torsional oscillations is as follows:

$$J = m\left(\frac{\varepsilon\omega_C^2}{\Omega_p^2 - \omega_C^2}\right)^2\left(1 + \frac{1}{A_\varphi}\right), \tag{4.133}$$

where m is the rotor mass, ε is the rotor mass eccentricity, Ω_p is the natural resonance frequency of the controlled object, ω_C is the mean angular velocity, and A_φ is the amplitude of the torsional oscillations.[8]

From reference data, it is known that the rotor mass is determined with high accuracy and is equal to (0.116 ± 0.0015) kg. It is also known that the relative uncertainty in measuring the maximum rotor mass eccentricity, which is $\varepsilon = 158$ μm, does not exceed $\delta_\varepsilon = 5\%$, and the natural resonance frequency of conical rotor systems with a gas-magnetic suspension can be (90 ± 10) Hz.

The experimental values of the SDs of the angular velocity measurement errors for repeated measurements and different sampling periods of the three observation groups are: $\sigma_{\omega 1} = 5.52\ \text{rad/s}, \sigma_{\omega 2} = 2.11\ \text{rad/s}, \text{and}\ \sigma_{\omega 3} = 1.72\ \text{rad/s}$.

Since we have three groups of repeated measurements, each with $K = 1000$ observations, the mean of the SDs is calculated using the following equation:

$$\bar{\sigma}_\omega = \frac{1}{N}\sum_{n=1}^{N} \sigma_{\omega n} = \frac{5.52 + 2.11 + 1.72}{3} = 3.11\ \text{rad/s}. \tag{4.134}$$

The Type A experimental standard uncertainty for the three groups of SDs is calculated using the following equation:

$$u_{A\omega} = \sqrt{\frac{1}{N-1}\sum_{n=1}^{N} (\sigma_{\omega n} - \bar{\sigma}_\omega)^2} = 2.09\ \text{rad/s}. \tag{4.135}$$

Next, we determine whether the between-group component of the variance is significant compared to the within-group component. For this, we calculate the first estimate, denoted as s_I^2, under the assumption that the between-group variance is zero:

$$s_I^2 = K u_{A\omega}^2 = \frac{K}{N-1}\sum_{n=1}^{N} (\sigma_{\omega n} - \bar{\sigma}_\omega)^2 = 4370.1\ (\text{rad/s})^2. \tag{4.136}$$

This estimate (4.136) has $v_I = N - 1$ degrees of freedom.

The second estimate, denoted as s_{II}^2, is the average of the variances obtained from $N = 3$ individual within-group variance values. Since the design is balanced and all degrees of freedom $v_n = K - 1$, the estimate s_{II}^2 is the mean $\sigma_{\omega n}{}^2(\sigma_{\omega k})$:

$$s_{II}^2 = \frac{1}{N}\sum_{n=1}^{N} \sigma_{\omega n}^2(\sigma_{\omega k}) = \frac{5.52^2 + 2.11^2 + 1.72^2}{3} = 12.63\ (\text{rad/s})^2. \tag{4.137}$$

The second estimate (4.137) has $v_{II} = N(K - 1)$ degrees of freedom.

To compare the estimates s_I^2 and s_{II}^2, the F-test is used. It is known that the F-distribution of the components is the probability distribution of the ratio:

$$F(v_I, v_{II}) = \frac{s_I^2(v_I)}{s_{II}^2(v_{II})} = \frac{4370.1}{12.63} = 346. \tag{4.138}$$

Since the estimate s_I^2 is greater than the estimate s_{II}^2, this indicates the presence of a between-group variance, which is assumed to be random. Then, the estimated variance $\bar{\sigma}_\omega$ (4.134) is obtained from the variance $u_{A\omega}^2$ (4.135), as it appropriately reflects both the within-group and between-group random components of the variance. Thus, the Type A standard uncertainty of the angular velocity measurement is determined using the following expression:

$$u_{A\omega}^2(\bar{\sigma}_\omega) = \frac{u_{A\omega}^2}{N} = 1.46\ \text{rad/s}. \tag{4.139}$$

From experimental data, it is known that the allowable deviations of the torsional oscillation amplitudes are $\Delta_A = \pm 8 \cdot 10^{-5}$ rad/s. The Type B standard uncertainty of the torsional oscillation measurement is calculated using the following equation:

$$u_{B_{A\varphi}} = \frac{\Delta_A}{\sqrt{3}} = \frac{8 \cdot 10^{-5}}{1.73} = 4.62 \cdot 10^{-5} \text{ rad/s}. \tag{4.140}$$

The Type B standard uncertainty due to the inexact measurement of the rotor mass, assuming a uniform probability distribution, is estimated using the following equation:

$$u_{Bm} = \frac{\Delta_m}{\sqrt{3}} = \frac{0.0015}{1.73} = 0.87 \cdot 10^{-3} \text{ kg}. \tag{4.141}$$

The Type B standard uncertainty due to deviations of the resonance frequency of the conical rotor system is estimated using the following equation:

$$u_{B\Omega} = \frac{\Delta_\Omega}{\sqrt{3}} = \frac{10}{1.73} = 5.78 \text{ Hz}. \tag{4.142}$$

The Type B standard uncertainty of the rotor mass eccentricity measurement is equal to:

$$u_{B\varepsilon} = \frac{\varepsilon}{100\%\sqrt{3}} \delta_\varepsilon = \frac{158}{173} 5 = 4.57 \text{ µm}. \tag{4.143}$$

The combined standard uncertainty of the indirect measurement of the rotor's moment of inertia, considering the correlation between angular velocity and torsional oscillations, is calculated using the following equation:

$$u_c^2(J) = \left(\frac{\partial J}{\partial \omega_C}\right)^2 u_{A\omega}^2(\bar{\sigma}_\omega) + \left(\frac{\partial J}{\partial A_\varphi}\right)^2 u_{B_{A\varphi}}^2 + \left(\frac{\partial J}{\partial m}\right)^2 u_{Bm}^2 +$$
$$\left(\frac{\partial J}{\partial \varepsilon}\right)^2 u_{B\varepsilon}^2 + \left(\frac{\partial J}{\partial \Omega_p}\right)^2 u_{B\Omega}^2, \tag{4.144}$$

where $\frac{\partial J}{\partial \omega_C} = 4\frac{m\varepsilon^2\omega_C^3}{\left(\Omega_p^2 - \omega_C^2\right)^2}\left(1 + \frac{1}{A_\varphi}\right) + 4\frac{m\varepsilon^2\omega_C^5}{\left(\Omega_p^2 - \omega_C^2\right)^3}\left(1 + \frac{1}{A_\varphi}\right)$ is the sensitivity coefficient with respect to angular velocity, equal to $-0.467 \cdot 10^{-10}$ kg·s·m^2/rad at a maximum angular velocity of 10^4 rad/s and a maximum torsional oscillation of $8 \cdot 10^{-5}$ rad/s, with the above-mentioned parameters entering the equation for determining the moment of inertia, $\frac{\partial J}{\partial A_\varphi} = -\frac{m\varepsilon^2\omega_C^4}{\left(\Omega_p^2 - \omega_C^2\right)^2 A_\varphi{}^2}$ is sensitivity coefficient with respect to torsional oscillations, equal to -0.455 kg·s·m^2/rad for the same rotor system parameters, $\frac{\partial J}{\partial m} = \frac{\varepsilon^2\omega_C^4}{\left(\Omega_p^2 - \omega_C^2\right)^2}\left(1 + \frac{1}{A_\varphi}\right)$ is sensitivity coefficient with respect to the rotor mass, equal to $0.31 \cdot 10^{-3}$ m^2 for the same rotor system parameters, $\frac{\partial J}{\partial \varepsilon} = 2\frac{m\varepsilon\omega_C^4}{\left(\Omega_p^2 - \omega_C^2\right)^2}\left(1 + \frac{1}{A_\varphi}\right)$ is sensitivity coefficient with respect to the rotor mass eccentricity, equal to 0.461 kg·m for the same rotor system parameters, and $\frac{\partial J}{\partial \Omega_p} = -4\frac{\Omega_p m\varepsilon^2\omega_C^4}{\left(\Omega_p^2 - \omega_C^2\right)^3}\left(1 + \frac{1}{A_\varphi}\right)$ is sensitivity coefficient with respect to the rotor mass eccentricity, equal to $0.83 \cdot 10^{-9}$ kg·s·m^2 for the same rotor system parameters.

Thus, substituting the corresponding sensitivity coefficients and evaluated uncertainties into Equation (4.144), we obtain the estimate of the combined standard uncertainty of the indirect measurement of the rotor moment of inertia, which equals $u_c(J) = 2.11 \cdot 10^{-5}$ kg·m^2.

The expanded uncertainty of the result of the indirect measurement of the rotor moment of inertia is determined using the following equation:

$$U = t(\nu_{\text{eff}})u_c(J) = 1.96 \cdot 2.11 \cdot 10^{-5} = \pm 4.14 \cdot 10^{-5}\ \text{kg} \cdot \text{m}^2 \text{ for } p = 0.95. \quad (4.145)$$

Thus, based on the measurements of angular velocity and the amplitude of torsional oscillations, the combined standard uncertainty of the indirect measurement of the rotor moment of inertia has been evaluated and is found not to exceed $2.11 \cdot 10^{-5}$ kg·m^2 with a confidence probability of 0.95, while the expanded uncertainty of the rotor moment of inertia measurement is equal to $\pm 4.14 \cdot 10^{-5}$ kg·m^2.

4.4.11 Evaluation of the Uncertainty of Collective Mass Measurement

As a result of weighing, the following combinations of the masses of two weights were obtained:

$$m1 = 4.97 \text{ kg};\ m2 = 1.02 \text{ kg};\ m1 + m2 = 6.08 \text{ kg};\ m1 - m2 = 4.02 \text{ kg}.$$

It is required to estimate the masses $M1$ and $M2$ of the weights and their uncertainties at a confidence level $p = 0.95$.

To estimate the masses of the weights, we compile a system of conditional equations based on the measured combinations of the masses, of the form:

$$\left.\begin{array}{l} 1 \cdot m_1 + 0 \cdot m_2 = 4.97; \\ 0 \cdot m_1 + 1 \cdot m_2 = 1.02; \\ 1 \cdot m_1 + 1 \cdot m_2 = 6.08; \\ 1 \cdot m_1 - 1 \cdot m_2 = 4.02. \end{array}\right\} \quad (4.146)$$

Next, we construct the system of normal equations. To do this, we algebraically sum the left- and right-hand sides of the system (4.146), first using the coefficients for the mass values $m1$ that are non-zero (i.e., that participate in the ith measurement), and then using the coefficients for the mass values $m2$ that are non-zero as well. As a result, we obtain:

$$\left.\begin{aligned} 3\cdot m_1 + 0\cdot m_2 &= 15.07; \\ 0\cdot m_1 + 3\cdot m_2 &= 3.08. \end{aligned}\right\} \tag{4.147}$$

By solving the system of normal equations (4.147), we obtain the estimate of the masses of the weights in the case of collective measurement: $M1 = 5.02$ kg; $M2 = 1.03$ kg.

The obtained estimates of the values of the masses of the weights $M1$ and $M2$ are substituted into the system of conditional Equations (4.146), which results in the residuals:

$$\left.\begin{aligned} \delta_1 &= 4.97 - 5.02 = -0.05 \text{ kg}; \\ \delta_2 &= 1.02 - 1.03 = -0.01 \text{ kg}; \\ \delta_3 &= 6.08 - (5.02 + 1.03) = 0.03 \text{ kg}; \\ \delta_4 &= 4.02 - (5.02 - 1.03) = 0.03 \text{ kg}. \end{aligned}\right\} \tag{4.148}$$

The numerical values of the obtained residuals (4.148) characterize the discrepancy (inconsistency) between the left- and right-hand sides of the system of conditional Equations (4.146).

The standard uncertainty of the conditional Equations (4.146) is calculated using the equation:

$$u(\delta) = \sqrt{\frac{\sum_{i=1}^{n}\delta_i^2}{n-q}} = \sqrt{\frac{(-0,05)^2 + (-0.01)^2 + 0.03^2 + 0.03^2}{4-2}} = 0.05 \text{ kg}, \tag{4.149}$$

where $n = 4$ is the number of combinations of measured values (conditional equations), $q = 2$ is the number of quantities (estimates) to be determined.

The standard uncertainty of the first estimate of the mass of weight $M1$ is calculated using the equation:

$$u(M1) = u(\delta)\sqrt{\frac{D_{11}}{D}} = 0.05\sqrt{\frac{3}{\begin{vmatrix} 3 & 0 \\ 0 & 3 \end{vmatrix}}} = 0.05\sqrt{\frac{3}{9}} = 0.03 \text{ kg}, \tag{4.150}$$

where D is the main determinant, D_{11} is the adjunct, obtained from the system of normal equations (4.147) by deleting the first row and the first column.

The standard uncertainty of the second estimate of the mass of the weight $M2$ is calculated using the equation:

$$u(M2) = u(\delta)\sqrt{\frac{D_{22}}{D}} = 0.05\sqrt{\frac{3}{\begin{vmatrix} 3 & 0 \\ 0 & 3 \end{vmatrix}}} = 0.05\sqrt{\frac{3}{9}} = 0.03 \text{ kg}, \tag{4.151}$$

where D_{22} is the cofactor (adjoint) determined from the system of normal equations (4.147) by deleting the second row and the second column.

The expanded uncertainties of the estimates of the masses of weights $M1$ and $M2$ at a confidence level of $p = 0.95$ are determined using the equations:

$$U_{M1} = t_p \cdot u(M2) = 4.3 \cdot 0.03 = 0.12\ \text{kg};\ U_{M2} = t_p \cdot u(M2) = 4.3 \cdot 0.03 = 0.12\ \text{kg}, \quad (4.152)$$

where t_p is Student's t coefficient for the number of degrees of freedom $v = n - q = 4 - 2 = 2$ and a confidence level $p = 0.95$ $\left(t_p = 4.3\right)$.

Thus, the result of the combined measurement of the weights' masses can be expressed as:

$M1 = 5.02 \pm 0.12$ kg for $p = 0.95$;
$M2 = 1.03 \pm 0.12$ kg for $p = 0.95$.

4.4.12 Evaluation of the Uncertainty of Concurrent Measurement Results

Consider the case in which we have a system of 9 equations of concurrent measurements:

$$\begin{cases} A_{11}x_1 + A_{12}x_2 + A_{13}x_3 + A_{14}x_4 = y_1, \\ A_{21}0 + A_{22}x_2 + A_{23}x_3 + A_{24}0 = y_2, \\ A_{31}x_1 + A_{32}x_2 + A_{33}0 + A_{34}0 = y_3, \\ A_{41}0 + A_{42}0 + A_{43}x_3 + A_{44}x_4 = y_4, \\ A_{51}x_1 + A_{52}x_2 + A_{53}x_3 + A_{54}0 = y_5, \\ A_{61}x_1 + A_{62}0 + A_{63}x_3 + A_{64}x_4 = y_6, \\ A_{71}0 + A_{72}x_2 + A_{73}x_3 + A_{74}x_4 = y_7, \\ A_{81}0 + A_{82}x_2 + A_{83}0 + A_{84}x_4 = y_8, \\ A_{91}x_1 + A_{92}0 + A_{93}x_3 + A_{94}0 = y_9. \end{cases} \quad (4.153)$$

For an illustrative example of solving such a measurement task, we shall use the values given in Table 4.7.

In accordance with the generally accepted methodology for evaluating the results of concurrent measurements, a system of normal equations is constructed, where the number of equations in the system must correspond to the number of unknown values of the quantities:

$$\begin{cases} B_{11}x_1 + B_{12}x_2 + B_{13}x_3 + B_{14}x_4 = c_1; \\ B_{21}x_1 + B_{22}x_2 + B_{23}x_3 + B_{24}x_4 = c_2; \\ B_{31}x_1 + B_{32}x_2 + B_{33}x_3 + B_{34}x_4 = c_3; \\ B_{41}x_1 + B_{42}x_2 + B_{43}x_3 + B_{44}x_4 = c_4, \end{cases} \quad (4.154)$$

where the corresponding coefficients are given by: $B_{11} = \sum_{j=1}^{9} A_{j1}^2 = 5$, $B_{22} = \sum_{j=1}^{9} A_{j2}^2 = 6$, $B_{33} = \sum_{j=1}^{9} A_{j3}^2 = 7$, $B_{44} = \sum_{j=1}^{9} A_{j4}^2 = 5$, $B_{12} = B_{21} = \sum_{j=1}^{9} A_{j1} \cdot A_{j2} = 3$, $B_{13} = B_{31} = \sum_{j=1}^{9} A_{j1} \cdot A_{j3} = 4$, $B_{14} = B_{41} = \sum_{j=1}^{9} A_{j1} \cdot A_{j4} = 2$, $B_{24} = B_{42} = \sum_{j=1}^{9} A_{j4} \cdot A_{j2} = 3$, $B_{23} = B_{32} = \sum_{j=1}^{9} A_{j3} \cdot A_{j2} = 4$, $B_{34} = B_{43} = \sum_{j=1}^{9} A_{j3} \cdot A_{j4} = 4$, $c_1 = \sum_{j=1}^{9} A_{j1} y_j = 14.13 + 7.17 + 10.13 + 12.26 + 8.15 = 51.84$, $c_2 = \sum_{j=1}^{9} A_{j2} y_j = 14.13 + 5.14 + 7.17 + 10.13 + 12.19 + 6.17 = 54.93$, $c_3 = \sum_{j=1}^{9} A_{j3} y_j = 14.13 + 5.14 + 7.15 + 10.13 + 12.26 + 12.19 + 8.15 = 69.15$, $c_4 = \sum_{j=1}^{9} A_{j4} y_j = 14.13 + 7.15 + 12.26 + 6.17 + 12.19 = 51.9$.

Considering the values of these coefficients, the system of normal equations (4.154) takes the form:

$$\begin{cases} 5x_1 + 3x_2 + 4x_3 + 2x_4 = 51.84; \\ 3x_1 + 6x_2 + 4x_3 + 3x_4 = 54.93; \\ 4x_1 + 4x_2 + 7x_3 + 4x_4 = 69.15; \\ 2x_1 + 3x_2 + 4x_3 + 5x_4 = 51.9. \end{cases} \quad (4.155)$$

The principal determinant of this system is given by the expression:

$$D = \begin{vmatrix} 5 & 3 & 4 & 2 \\ 3 & 6 & 4 & 3 \\ 4 & 4 & 7 & 4 \\ 2 & 3 & 4 & 5 \end{vmatrix} = 168. \quad (4.156)$$

The measured quantities are evaluated by the expressions:

$$\bar{x}_1 = \frac{D_{x1}}{D}, \quad \bar{x}_2 = \frac{D_{x2}}{D}, \quad \ldots, \quad \bar{x}_m = \frac{D_{xm}}{D}. \quad (4.157)$$

The determinants D_{x1}, D_{x2}, D_{x3}, and D_{x4} are obtained from the principal determinant D (4.156) by replacing the column of coefficients corresponding to the unknown x_m with the column of free terms c_j:

$$D_{x1} = \begin{vmatrix} 51.84 & 3 & 4 & 2 \\ 54.93 & 6 & 4 & 3 \\ 69.15 & 4 & 7 & 4 \\ 51.9 & 3 & 4 & 5 \end{vmatrix} = 730.83;$$

$$D_{x2} = \begin{vmatrix} 5 & 51.84 & 4 & 2 \\ 3 & 54.93 & 4 & 3 \\ 4 & 69.15 & 7 & 4 \\ 2 & 51.9 & 4 & 5 \end{vmatrix} = 415.33;$$

$$D_{x3} = \begin{vmatrix} 5 & 3 & 51.9 & 2 \\ 3 & 6 & 54.93 & 3 \\ 4 & 4 & 69.15 & 4 \\ 2 & 3 & 51.9 & 5 \end{vmatrix} = 585;$$

$$D_{x4} = \begin{vmatrix} 5 & 3 & 4 & 51.84 \\ 3 & 6 & 4 & 54.93 \\ 4 & 4 & 7 & 69.15 \\ 2 & 3 & 4 & 51.9 \end{vmatrix} = 734.19.$$

Considering expressions (4.157), the most probable values of the unknown quantities are:

$$x_1 = 730.83/168 = 4.35;$$

$$x_2 = 415.53/168 = 2.47;$$

$$x_3 = 585/168 = 3.48;$$

$$x_4 = 734.19/168 = 4.37.$$

By substituting the most probable values into the conditional equations, we obtain:

$$x_1 + x_2 + x_3 + x_4 = 4.35 + 2.47 + 3.48 + 4.37 = 14.67;$$

$$x_2 + x_3 = 2.47 + 3.48 = 5.95;$$

$$x_1 + x_2 = 4.35 + 4.37 = 8.72;$$

$$x_3 + x_4 = 3.48 + 4.37 = 7.85;$$

$$x_1 + x_2 + x_3 = 4.35 + 2.47 + 3.48 = 10.3;$$

$$x_1 + x_3 + x_4 = 4.35 + 3.48 + 4.37 = 12.2;$$

$$x_2 + x_4 + x_3 = 2.47 + 4.37 + 3.48 = 10.32;$$

$$x_2 + x_4 = 2.47 + 4.37 = 6.84;$$

$$x_3 + x_1 = 3.48 + 4.35 = 7.83.$$

After evaluating the values of the measured quantities, the residuals are determined, representing the differences between the results of the concurrent measurements (Table 4.7) and the obtained estimates:

TABLE 4.7
Results of Concurrent Measurements

$x_1 + x_2 + x_3 + x_4$	$x_2 + x_3$	$x_1 + x_2$	$x_3 + x_4$	$x_1 + x_2 + x_3$	$x_1 + x_3 + x_4$	$x_2 + x_4 + x_3$	$x_2 + x_4$	$x_3 + x_1$	P
14.13	5.14	7.17	7.15	10.13	12.26	12.19	6.17	8.15	0.95

$$v_1 = 14.13 - 14.67 = -0.54;$$

$$v_2 = 5.14 - 5.95 = -0.81;$$

$$v_3 = 7.17 - 8.72 = -1.55;$$

$$v_4 = 7.15 - 7.85 = -0.7;$$

$$v_5 = 10.13 - 10.3 = -0.17;$$

$$v_6 = 12.26 - 12.2 = 0.06;$$

$$v_7 = 12.19 - 10.32 = 1.87;$$

$$v_8 = 6.17 - 6.84 = -0.67;$$

$$v_9 = 8.15 - 7.83 = 0.32.$$

After determining the residuals, the uncertainties of the measured quantities are calculated using the expression:

$$u_{x1} = \sqrt{\frac{A_{11}}{D} \cdot \frac{\sum_{j=1}^{m} v_j^2}{m-n}}. \tag{4.158}$$

Similarly, for the other measured quantities, the uncertainties u_{x_2}, u_{x_3}, and u_{x_4} are determined. However, before this, the values of the cofactors must be calculated. These are obtained from the main determinant D by removing the nth column and the mth row corresponding to the measured quantity being evaluated, followed by multiplication by $(-1)^{n+m}$. That is, to calculate the cofactor A_{11}, the first column and first row are removed from the principal determinant, and the resulting determinant is computed:

$$A_{11} = \begin{vmatrix} 6 & 4 & 3 \\ 4 & 7 & 4 \\ 3 & 4 & 5 \end{vmatrix} = 67.$$

To determine the cofactor A_{22}, the second column and the second row are removed from the principal determinant D, and the resulting determinant is computed:

$$A_{22} = \begin{vmatrix} 5 & 4 & 2 \\ 4 & 7 & 4 \\ 2 & 4 & 5 \end{vmatrix} = 51.$$

To determine the cofactor A_{33}, the third column and the third row are removed from the principal determinant D, and the resulting determinant is computed:

$$A_{33} = \begin{vmatrix} 5 & 3 & 2 \\ 3 & 6 & 3 \\ 2 & 3 & 5 \end{vmatrix} = 72.$$

To determine the cofactor A_{44}, the third column and third row are removed from the principal determinant D, and the resulting determinant is calculated:

$$A_{44} = \begin{vmatrix} 5 & 3 & 4 \\ 3 & 6 & 4 \\ 4 & 4 & 7 \end{vmatrix} = 67.$$

Thus, the values of the uncertainties of the concurrent measurements are obtained using Equation (4.158):

$$u_{x_1} = 0.79;$$

$$u_{x_2} = 0.69;$$

$$u_{x_3} = 0.82;$$

$$u_{x_4} = 0.79.$$

For degrees of freedom $m - n = 5$ and a probability $p = 0.95$, Student's coefficient is $t_p = 2.57$. Taking this value into account, the expanded uncertainties of the combined measurements are:

$$U_1 = \pm 2.57 \cdot 0.79 = \pm 2.03;$$

$$U_2 = \pm 2.57 \cdot 0.69 = \pm 1.77;$$

$$U_3 = \pm 2.57 \cdot 0.82 = \pm 2.11;$$

$$U_4 = \pm 2.57 \cdot 0.79 = \pm 2.03.$$

Therefore, the results of the concurrent measurement can be presented in the form:

$$x_1 = 4.35 \pm 2.03, p = 0.95;$$

$$x_2 = 2.47 \pm 1.77, p = 0.95;$$

$$x_3 = 3.48 \pm 2.11, p = 0.95;$$

$$x_4 = 4.37 \pm 2.03, p = 0.95.$$

4.4.13 Quantitative Evaluation of the Emissivity of 316L Stainless Steel Powder

When determining the emissivity of 316L powder steel, the following instruments were used: the IR camera FLIR A700, a device for measuring relative humidity, a device for measuring ambient temperature, a tape measure for measuring the distance from the IR camera to the measurement object, precision channels for measuring temperature with thermocouples. Each of the listed instruments contributes to the combined measurement uncertainty, and therefore, it is necessary to estimate the measurement uncertainty of each of them.[9]

Based on the obtained results of experimental studies, it is possible to estimate the standard measurement uncertainty of Type A $u_{Aj}(\overline{x})$ for each reference temperature value using the expression:

$$u_{Aj}(\overline{x}) = \left[\frac{\sum_{i=1}^{n} (x_i - \overline{x})^2}{n(n-1)} \right]^{0.5}, \tag{4.159}$$

where x_i is the quantity of measurements, $\overline{x}$ is the mean value, and n is the number of values in the sample.

The values of Type A standard uncertainties calculated using formula (4.159) for the old and virgin samples of 316L powder steel at fixed reference temperature values are shown in Table 4.8.

Based on the values of Type A uncertainties of temperature measurements calculated by the formula (4.159) at each reference point, we calculate the combined value of Type A uncertainty according to the equation:

$$u_{Ac} = \left[\frac{\sum_{j=1}^{m} u_{Aj}^2(\overline{x})}{G} \right]^{0.5}, \tag{4.160}$$

where G is the number of investigated groups of arrays ($G = 26$).

Substituting into Equation (4.160) the values of Type A uncertainties obtained above at each reference point, we obtained a combined Type A uncertainty value of $u_{Ac} = 1.47°C$.

The relative values of the combined measurement uncertainty of Type A for the lower and upper ranges of changes in the temperature of the object of study will be calculated using the equation:

TABLE 4.8
The Values of Type A Uncertainties at Fixed Reference Temperature Values

The Reference Temperature of the Old Sample, °C	The Reference Temperature of the Virgin Sample, °C	Type A Uncertainty for the Old Sample, °C	Type A Uncertainty for the Virgin Sample, °C
59	63	1.62	1.51
62	67	1.2	1.91
100	103.5	0.79	1.39
102	105.5	0.98	1.24
352	365	1.72	2.31
362	378	1.66	2.23
400	409	1.12	1.21
415	424	1.21	1.53
557	569	1.06	1.34
800	817	1.45	1.62
834	847	1.25	1.28
893	901	1.35	1.58
898	905	1.15	1.38

$$\tilde{u}_{Ac_l} = \frac{u_{Ac}}{|t_{ob}|} 100\%, \tag{4.161}$$

where t_{ob} is the object temperature.

Substituting into Equation (4.161) the calculated values of the combined uncertainties of Type A and the temperature of the research object at reference points (see Table 4.8.) we obtained the following values of the relative combined uncertainties of Type A $\tilde{u}_{Ac_{59}} = 2.49\%$, $\tilde{u}_{Ac_{102}} = 1.44\%$, $\tilde{u}_{Ac365} = 0.4\%$, $\tilde{u}_{Ac_{412}} = 0.36\%$, $\tilde{u}_{Ac_{563}} = 0.26\%$, $\tilde{u}_{Ac_{825}} = 0.18\%$, and $\tilde{u}_{Ac_{900}} = 0.16\%$.

At the next stage, it is necessary to evaluate the uncertainties that are introduced due to the limited properties of the equipment used.

Since we know from the calibration certificate the absolute error of the precision temperature MC, which is $\Delta_{\text{ref}} = \pm\ 1.1°\text{C}$ in the temperature measurement range from 0 to 1260°C, based on this information we can estimate the Type B uncertainty, which is due to limited technical properties thermocouple.

Following the international recommendations ISO/IEC Guide 98-3:2008, if the type of error distribution law is not specified, then it is recommended to calculate the measurement uncertainty by Type B for a uniform distribution law. With this in mind, we calculated the measurement uncertainty of Type B, which is introduced by the precision temperature measuring channel, assuming a uniform distribution according to the equation:

$$u_{B\text{ref}} = \frac{\pm\ \Delta_{\text{ref}}}{\sqrt{3}}. \tag{4.162}$$

Substituting values of the absolute error $\Delta_{\text{ref}} = \pm 1.1°C$ in Equation (4.162), we obtain the values of the measurement uncertainty of the reference temperature, which is equal to $u_{B\text{ref}} = \pm 0.64°C$ in the measurement range from 0 to 1260°C.

Relative humidity measurements were determined using a portable device *ThermoPro* that has the absolute error of $\Delta_{\text{hum}} = 2\%$. Uncertainty of Type B of moisture measurement, assuming a normal distribution of errors, we calculate by the equation:

$$u_{B\text{hum}} = \frac{\Delta_{\text{hum}}}{k_p}, \tag{4.163}$$

where k_p is the coverage coefficient for a normal distribution with probability p, the value of which is taken from Student's t-distribution table and for probability $p = 0.95$ takes the value $k_p = 1.96$.

Substituting in Equation (4.163) the value of the absolute error $\Delta_{\text{hum}} = 2\%$ of relative humidity measurement, we obtain the value of the Type B uncertainty of relative humidity measurement, which is equal to $u_{B\text{hum}} = \frac{2}{1.96} = 1.02\%$ and is caused by the limited accuracy of the instrument for measuring humidity, assuming a normal distribution law with a confidence probability of 0.95.

The absolute error of the device *ThermoPro* for measuring the ambient temperature is $\Delta_{ta} = \pm 0.556°C$. Since the ambient temperature in the laboratory was measured using the same universal device that measured the relative humidity, then the uncertainty of Type B in measuring the ambient temperature is calculated by Equation (4.163). Substituting the value of the error $\Delta_{ta} = \pm 0.556°C$ in measuring the ambient temperature in Equation (4.163), we obtain the uncertainty in measuring the ambient temperature, which is equal to $u_{Bta} = \frac{0.556}{1.96} = 0.28°C$.

The distance from the camera to the object was measured with a tape measure with an absolute error of $\Delta_d = \pm 1$mm. Since the law of error distribution is not known, in accordance with international recommendations, we accept its type as close to uniform and determine the uncertainty of distance measurement by the equation:

$$u_{Bd} = \frac{\pm \Delta_d}{\sqrt{3}}. \tag{4.164}$$

Substituting in Equation (4.164) the values of the absolute error of distance measurement $\Delta_d = 1$ mm, we obtain the values of the uncertainty of distance measurement under the assumption of a uniform distribution law, which is equal to $u_{Bd} = 1/1.73 = 0.58$ mm.

From the calibration certificate of the FLIR A700 infrared camera, it is known that the absolute error of temperature measurement is $\Delta_{\text{cam}} = \pm 2°C$ in the range from −20 to 2000°C. Based on the calibration results of the infrared camera, we will calculate the Type B uncertainty value of the temperature measurement that the camera introduces due to its limited properties, under the assumption of a uniform error distribution law, according:

$$u_{B\text{cam}} = \frac{\pm \Delta_{\text{cam}}}{\sqrt{3}} = \frac{\pm 2}{1.73} = \pm 1.16°C. \tag{4.165}$$

Since many additional instruments are used for experimental studies of the temperature and emissivity of the measurement object (316L powder steel), and each of them has different physical quantities (not only temperature, but also distance and humidity) and contributes to the overall measurement uncertainty, then for calculation of the total contribution to the uncertainty of measurements of Type B, it is advisable to proceed to the determination of the relative values of uncertainty and calculate the combined uncertainty of Type B $\tilde{u}_{Bc_{\text{inst}}}$ according to the equation:

$$\tilde{u}_{Bc_{\text{inst}}} = \left[\left(\frac{u_{B\text{ref}}}{|t_{\text{ref}}|}\right)^2 + \left(\frac{u_{B\text{hum}}}{|\gamma|}\right)^2 + \left(\frac{u_{Bta}}{|t_a|}\right)^2 + \left(\frac{u_{Bd}}{|d|}\right)^2 + \left(\frac{u_{B\text{cam}}}{|t_{ob}|}\right)^2\right]^{0.5}, \quad (4.166)$$

where t_{ref} is an estimate of the value of the reference temperature, which is obtained using a precision temperature measuring channel with a thermocouple. The relative measurement uncertainty of the reference temperature is calculated by the equation:

$$\tilde{u}_{B\text{ref}} = \frac{u_{B\text{ref}}}{|t_{\text{ref}}|}100\%, \quad (4.167)$$

and its value in the temperature measurement range from 59 to 900°C are $\tilde{u}_{B\text{ref}_{59}} = \frac{0.64}{59}100\% = 1.1\%$, $\tilde{u}_{B\text{ref}_{102}} = \frac{0.64}{102}100\% = 0.63\%$, $\tilde{u}_{B\text{ref}_{365}} = \frac{0.64}{365}100\% = 0.18\%$, $\tilde{u}_{B\text{ref}_{412}} = \frac{0.64}{412}100\% = 0.16\%$, $\tilde{u}_{B\text{ref}_{563}} = \frac{0.64}{563}100\% = 0.11\%$, $\tilde{u}_{B\text{ref}_{825}} = \frac{0.64}{825}100\% = 0.08\%$, $\tilde{u}_{B\text{ref}_{900}} = \frac{0.64}{900}100\% = 0.07\%$

γ is the relative humidity value in the laboratory. Since during the experiments the value of relative humidity was 47%, then considering this value of relative humidity, the relative uncertainty of humidity measurement will be

$$\tilde{u}_{B\text{hum}} = \frac{u_{B\text{hum}}}{|\gamma|}100\% = \frac{1.02}{47}100\% = 2.17\%; \quad (4.168)$$

t_a is the ambient temperature in the laboratory, which at the time of testing was measured by the combined device and amounted to 23°C. Considering this value of the ambient temperature, the relative uncertainty in the measurement of the ambient temperature, which is introduced due to the limited accuracy of the combined device used, is calculated by the equation:

$$\tilde{u}_{Bta} = \frac{u_{Bta}}{|t_a|}100\% = \frac{0.28}{23}100\% = 1.22\%; \quad (4.169)$$

d is the distance from the infrared camera to the measurement object. At the time of testing, the distance from the camera to the object of measurement was measured with a tape measure and amounted to 0.58 m, while the value of the relative uncertainty in measuring the distance, which is introduced due to the limited accuracy characteristics of the tape measure, is determined by the equation:

$$\tilde{u}_{Bd} = \frac{u_{Bd}}{|d|}100\% = \frac{0.58 \cdot 10^{-3}}{0.58}100\% = 0.1\%; \quad (4.170)$$

t_{ob} is the temperature of the object of study (316L powder steel). The value of the relative uncertainty of Type B of the measurement of the temperature of an object, which arises through the limited properties of the infrared camera, taking into account the measured temperature value, is calculated by the equation:

$$\tilde{u}_{B\text{cam}} = \frac{u_{B\text{cam}}}{|t_{ob}|} 100\%. \tag{4.171}$$

Since the temperature of the powder steel during the research varied from 59 to 900°C, then substituting the minimum temperature value of 59°C of the object under study into Equation (4.171), we obtain the values of the relative uncertainty of Type B, which is equal to $\tilde{u}_{B\text{cam}59} = \frac{1.16}{59} 100\% = 1.97\%$. Also substituting into Equation (4.171) other reference values of the temperature under study, we obtain the corresponding values of the relative uncertainty of Type B, which will be equal to $\tilde{u}_{B\text{cam}102} = \frac{1.16}{102} 100\% = 1.14\%$, $\tilde{u}_{B\text{cam}365} = \frac{1.16}{365} 100\% = 0.32\%$, $\tilde{u}_{B\text{cam}412} = \frac{1.16}{412} 100\% = 0.28\%$, $\tilde{u}_{B\text{cam}563} = \frac{1.16}{563} 100\% = 0.21\%$, $\tilde{u}_{B\text{cam}825} = \frac{1.16}{825} 100\% = 0.14\%$, and $\tilde{u}_{B\text{cam}900} = \frac{1.16}{900} 100\% = 0.13\%$.

As shown by the results of calculating the Type B relative uncertainty, which arises due to the limited accuracy characteristics of the infrared camera, its maximum value of $\tilde{u}_{B\text{cam}59} = 1.97\%$ occurs when measuring small values of the investigated powder steel temperature.

Thus, substituting in Equation (4.166) the values of relative uncertainties of Type B calculated according to formulas (4.167)–(4.171), each of which contributes to the combined (total) uncertainty of Type B, which arises due to the limited properties of the instruments used in research, we got: $\tilde{u}_{Bc_{\text{inst}59}} = 3.4\%$, $\tilde{u}_{Bc_{\text{inst}102}} = 2.81\%$, $\tilde{u}_{Bc_{\text{inst}365}} = 2.52\%$, $\tilde{u}_{Bc_{\text{inst}412}} = 2.51\%$, $\tilde{u}_{Bc_{\text{inst}563}} = 2.5\%$, $\tilde{u}_{Bc_{\text{inst}825}} = 2.5\%$, $\tilde{u}_{Bc_{\text{inst}900}} = 2.5\%$.

The temperature measurement uncertainty that occurs when choosing the emissivity by $\Delta\varepsilon = 0.01$ more or less than its actual value, assuming its normal distribution law, for the lower and upper values the temperature of the object was calculated using the equation:

$$u_{Be} = \frac{\Delta t}{k_p}. \tag{4.172}$$

The maximum deviation that occurs when the erroneous choice of emissivity by 0.01 comes from the results of the experiments presented in Table 4.8 (for virgin metal powder) at the reference temperature of 63°C. Taking into account this value of the maximum difference ($\Delta t = 65.75 - 61.4 = 4.35$°C), we calculate by formula (4.172) the maximum value of the temperature measurement uncertainty that can occur when the emissivity is inaccurately determined by 0.01 at the confidence level of 95% $u_{Be63} = \frac{4.35}{1.96} = 2.22$°C.

When measuring a temperature of about 102°C, the maximum value of the measurement uncertainty caused by an incorrect choice of the emissivity by 0.01 was $u_{Be102} = \frac{102.23-99.14}{1.96} = 1.58$°C. When measuring a temperature of about 352°C, the maximum value of the measurement uncertainty caused by an incorrect choice of the emissivity by 0.01 was $u_{Be352} = \frac{350.27-346.22}{1.96} = 2.1°C$. When measuring a temperature of about 400°C, the maximum value of the measurement uncertainty caused by an incorrect choice of the emissivity by 0.01 was $u_{Be400} = \frac{399.92-396.03}{1.96} = 1.98u_{Be352} = \frac{399.92-396.03}{1.96} = 1.98°C$. When measuring a temperature of about 557°C, the maximum value of the measurement uncertainty caused by an incorrect choice of the emissivity by 0.01 was $u_{Be557} = \frac{557.01-556.24}{1.96} = 0.4°C$. When measuring a temperature of about 817°C, the maximum value of the measurement uncertainty caused by an incorrect choice of the emissivity by 0.01 was $u_{Be817} = \frac{817.8-817.01}{1.96} = 0.4°C$. When measuring a temperature of about 900°C, the maximum value of the measurement uncertainty caused by an incorrect choice of the emissivity by 0.01 was $u_{Be901} = \frac{901.84-900.63}{1.96} = 0.62°C$.

Calculations show that the maximum value of the measurement uncertainty caused by the wrong choice of the emissivity by 0.01 when measuring the temperature at the lower limit of 63°C did not exceed $u_{Be63} = 2.22°C$.

The relative uncertainty of temperature measurements, which was caused by an incorrect choice of emissivity by 0.01, was calculated by the equation:

$$\tilde{u}_{Be_l} = \frac{u_{Be_l}}{|t_{ob}|} 100\%. \tag{4.173}$$

Substituting into Equation (4.173), the maximum values of the temperature measurement uncertainties calculated above at each reference point of the study, caused by the wrong choice of the emissivity by 0.01, we obtained $\tilde{u}_{Be63} = \frac{2.22}{63} 100\% = 3.52\%$, $\tilde{u}_{Be102} = \frac{1.58}{102} 100\% = 1.55\%$, $\tilde{u}_{Be352} = \frac{2.1}{352} 100\% = 0.6\%$, $\tilde{u}_{Be400} = \frac{1.98}{400} 100\% = 0.5\%$, $\tilde{u}_{Be557} = \frac{0.4}{557} 100\% = 0.07\%$, $\tilde{u}_{Be817} = \frac{0.4}{817} 100\% = 0.05\%$, and $\tilde{u}_{Be901} = \frac{0.62}{901} 100\% = 0.07\%$.

Based on the calculated values of the relative uncertainties of Type B measurements (4.166) and (4.173), as well as the relative combined uncertainty of Type A (4.160), we calculated the relative combined uncertainty of the temperature measurement results by the equation:

$$\tilde{u}_c = \left[\left(\tilde{u}_{Bc_{inst}}\right)^2 + \left(\tilde{u}_{Be_l}\right)^2 + \left(\tilde{u}_{Acl}\right)^2 \right]^{0.5}. \tag{4.174}$$

Substituting in Equation (4.174) the values of the relative uncertainties obtained for each of the reference points, we obtained $\tilde{u}_{c59} = \left[3.4^2 + 3.52^2 + 2.49^2\right]^{0.5} = 5.5\%$, $\tilde{u}_{c102} = \left[2.81^2 + 1.55^2 + 1.44^2\right]^{0.5} = 3.52\%$, $\tilde{u}_{c365} = \left[2.52^2 + 0.6^2 + 0.4^2\right]^{0.5} = 2.62\%$, $\tilde{u}_{c412} = \left[2.51^2 + 0.5^2 + 0.36^2\right]^{0.5} = 2.58\%$, $\tilde{u}_{c563} = \left[2.5^2 + 0.07^2 + 0.26^2\right]^{0.5} = 2.51\%$, $\tilde{u}_{c825} = \left[2.5^2 + 0.05^2 + 0.18^2\right]^{0.5} = 2.51\%$, and $\tilde{u}_{c900} = \left[2.5^2 + 0.07^2 + 0.16^2\right]^{0.5} = 2.51\%$.

To calculate the expanded uncertainty of temperature measurements, we recalculate the relative combined uncertainty (4.174) into absolute values of the combined uncertainty of temperature measurement for the lower and upper measurement ranges, as the result we got: assessment of the combined uncertainty of temperature measurement for the lower measurement range was $u_{c59} = \frac{t_{ob}}{100\%}\tilde{u}_{c59} = \frac{59°C}{100\%}5.5\% = 3.25°C$; assessment of the combined uncertainty of temperature measurement at the reference value of 102°C $u_{c102} = 3.59°C$; assessment of the combined uncertainty of temperature measurement at the reference value of 365°C $u_{c365} = 9.56°C$; assessment of the combined uncertainty of temperature measurement at the reference value of 412°C $u_{c412} = 10.63°C$; assessment of the combined uncertainty of temperature measurement at the reference value of 563°C $u_{c563} = 14.13°C$; assessment of the combined uncertainty of temperature measurement at the reference value of 825°C $u_{c825} = 20.71°C$; assessment of the combined uncertainty of temperature measurement for the upper measurement range was $u_{c900} = \frac{t_{ob}}{100\%}\tilde{u}_{c900} = \frac{900°C}{100\%}2.51\% = 22.59°C$.

The expanded uncertainty of temperature measurement was calculated by the equation:

$$U_j = t_p(\vartheta_{\text{eff}})u_c, \tag{4.175}$$

where $t_p(\vartheta_{\text{eff}})$ is Student's distribution coefficient for the coverage probability p and an effective degrees of freedom ϑ_{eff}.

The effective degrees of freedom are calculated using the Welch-Satterthwaite equation:

$$\vartheta_{\text{eff}} = (k-1)\frac{u_c^4}{u_{Ac}^4}. \tag{4.176}$$

Since each temperature measurement at one reference point was repeated at least $k = 6$ times, then substituting in Equation (4.176) the maximum value of the combined uncertainty and the combined uncertainty of Type A, we obtain $\vartheta_{\text{effmin}} = (6-1)\frac{3.25^4}{1.47^4} = 119$ and $\partial_{\text{eff}_{\text{max}}} = (6-1)\frac{22.59^4}{1.47^4} = 390386$.

In accordance with, if the effective degrees of freedom are more than 30, then they are considered to be sufficiently large and, in accordance with Student's table, with a probability $p = 0.95$, the coverage factor is assumed to be $t_p(\vartheta_{\text{eff}}) = 1.96$.

Then, substituting in Equation (4.175), the corresponding values for each of the reference points, we obtained the value of the expanded uncertainty of the results of temperature measurements: $U_{59} = \pm 6.37°C$; $U_{102} = \pm 7.04°C$; $U_{365} = \pm 18.74°C$; $U_{412} = \pm 20.83°C$; $U_{563} = \pm 27.69°C$; $U_{825} = \pm 40.59°C$; $U_{900} = 1.96*22.59 = \pm 44.28°C$.

Thus, the results of temperature measurements using the FLIR F700 infrared camera when 316L powder steel is heated to reference temperatures can be represented as: $T_{59} = 59 \pm 6.37°C, T_{102} = 102 \pm 7.04°C, T_{365} = 365 \pm 18.74°C, T_{412} = 412 \pm 20.83°C, T_{563} = 563 \pm 27.69°C, T_{825} = 825 \pm 40.59°C$, and $T_{900} = 900 \pm 44.28°C$ for $p = 0.95$.

Thus, the maximum relative combined uncertainty of temperature measurement of 316L powder steel was 5.5% in the measurement range from 59 to 900°C. Based on this maximum relative combined uncertainty in temperature measurement (5.5%), the potential combined uncertainty of the emissivity determination was analyzed. Calculating 5.5% of the maximum emissivity value of $\varepsilon = 0.58$, we obtained the maximum potential combined uncertainty in determining the emissivity of 316L powder steel, which was $u_{c\varepsilon} = 0.58 \cdot 0.055 = 0.03$. Then, using formula (4.175), we calculated the potential value of the expanded uncertainty in determining the emissivity, which was $U_\varepsilon = \pm 1.96 \cdot 0.03 = \pm 0.06$.

As can be seen from Figure 4.5, the emissivity of 316L powder steel varied from 0.31 to 0.58 when the temperature of the metal powder changed (increased) from 59 to 900°C. At the same time, from the experimental data it is clear (Figure 4.5) that in the temperature range from 59 to 370°C, the average value of the emissivity was $\varepsilon = 0.33 \pm 0.06$. For the temperature range from 370 to 470°C, the average emissivity value is $\varepsilon = 0.42 \pm 0.06$. For the temperature range from 470 to 900°C, the average emissivity value is $\varepsilon = 0.55 \pm 0.06$. The increase in emissivity at higher temperatures can be explained by oxidation effects at temperatures above 580°C. Since, as is known from literary sources,[10,11] the oxidation rate of metals at temperatures of approximately 580°C and especially 600°C increases greatly, which leads to an increase in the emissivity value. Therefore, when conducting this type of research, it is recommended to replace the metal powder sample with a fresh one after passing through a temperature barrier of 580°C.

To compare our results of measuring the emissivity of 316L powder steel and the measurement uncertainty with the experimental results obtained by other researchers, which are presented in, we used the E_n criterion (a measure of agreement between participants) by which we calculated the normalized deviation:

$$E_n = \frac{\left|\overline{\varepsilon}_{\text{exp}} - \overline{\varepsilon}_{\text{other}}\right|}{\left(U_{\text{exp}}^2 + U_{\text{other}}^2\right)^{0.5}}, \tag{4.177}$$

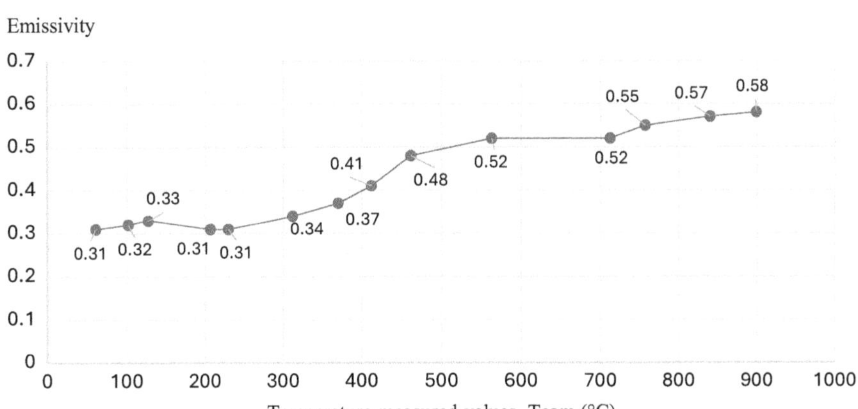

FIGURE 4.5 Characteristics of changes in the emissivity when the temperature of 316L powder steel changes from 59 to 900°C.

where $\bar{\varepsilon}_{exp}$ is the average value of the emissivity of powder steel 316L obtained in our laboratory, $\bar{\varepsilon}_{other}$ is the average value of the emissivity of 316L powder steel obtained by other researchers, the results of which are presented in, U_{exp} is the experimental value of the expanded uncertainty of measurement, calculated in our laboratory ($U_{exp} = \pm 0.06$), and U_{other} is the value of the expanded uncertainty of emissivity measurement, calculated by other researchers, the results of which are presented in ($U_{other} = \pm 0.05$).

As indicated in, the emissivity value of 316L powder steel varied from $\bar{\varepsilon}_{other1} = 0.37$ to $\bar{\varepsilon}_{other2} = 0.4$ in the temperature range from 200 to 500°C and from $\bar{\varepsilon}_{other2} = 0.4$ to $\bar{\varepsilon}_{other3} = 0.45$ in the temperature range from 500 to 600°C.

In our laboratory, we obtained experimental emissivity values for 316L powder steel, which varied from $\bar{\varepsilon}_{exp\,1} = 0.33$ to $\bar{\varepsilon}_{exp\,2} = 0.42$ in the temperature range from 59 to 370°C and from $\bar{\varepsilon}_{exp\,2} = 0.42$ to $\bar{\varepsilon}_{exp\,3} = 0.52$ in the temperature range from 370 to 700°C.

Substituting the known three pairs of emissivity values for 316L powder steel into Equation (4.177), we calculated three values for the measures of agreement (normalized deviation) between the results obtained by our laboratory and another laboratory, which were $E_{n1} = 0.51$, $E_{n2} = 0.26$, and $E_{n3} = \frac{|0.52-0.45|}{\left(0.06^2 + 0.05^2\right)^{0.5}} = 0.89$.

From the calculation results, it is clear that all measures of agreement of the results obtained do not exceed 1 ($E_n < 1$), which indicates that our laboratory is quite competent and no corrective actions are required.[12,13]

When the resultant E_n for any measurement is between +1 and −1 no corrective action is required. If the resultant E_n is greater than +1 or −1 this provides a signal that something might be wrong with a laboratory's calibration process, equipment, standards competence of personnel, stated uncertainties, etc. Participating laboratories can use such intercomparison: for cross training their personnel; demonstrating technical competence; and identifying potential areas for improvement.

4.4.14 Quantitative Evaluation of Measurement Uncertainty in Studies of Ionic Current Rectification

As a result of the experiments, current values $I_{i+} = [15, 10, 19, 7\ \text{nA}]$ were obtained at a positive potential of $\varphi_+ = 0.4$ V and current values $I_{i-} = [-15, -14, -21, -10\ \text{nA}]$ at a negative potential $\varphi_- = -0.4$ V.

Based on the obtained current values, Type A standard uncertainty is determined separately for positive and negative potentials using the equation:

$$u_A = \left[\frac{\sum_i^n \left(I_i - \bar{I}\right)^2}{n(n-1)}\right]^{0.5}, \tag{4.178}$$

where n is the number of measurements taken, and $\bar{I}$ is average current value.

The standard uncertainty of Type A calculated according to formula (4.178) at a positive potential of +0.4 V was $u_{AI+} = 2.66$ nA, and at a negative potential (−0.4 V) it was $u_{AI-} = 2.27$ nA.

When manufacturing a measuring needle from glass using a *Micrometer Setting P-2000*, the uncertainty arises, which, unfortunately, is very difficult to quantify. The suitability or unsuitability of the measuring head can be assessed using a microscope only by the experimenter by its geometrical form. But in the laboratory, there are results of studies of the obtained needle radii when they are created using the above installation. As a result of manufacturing the needles, it was found that the average value of the radii of the needles is $\overline{R} = 45.78$ nm, and the RMS deviation between the radii of the fabricated needles is $u_{BR} = \pm 0.75$ nm.

The uncertainty of the AgCl electrode potential, considering its technical characteristics, which indicate that the permissible potential deviation is $\Delta_1 = 3$ mV for a uniform distribution law, is determined by the equation:

$$u_{B1} = \frac{\Delta_1}{\sqrt{3}} = \frac{3\text{ mV}}{1.73} = 1.73\text{ mV}. \tag{4.179}$$

Type B uncertainty of preparing a standard solution of 100 mM *KCl* with a concentration of 7 pH or for other numerical values, which is carried out using the instrument *Accumet AB150*, can be calculated by the equation:

$$u_{B2} = \frac{\Delta_2}{\sqrt{3}} = \frac{0.01}{1.73} = 0.0058\text{ pH}, \tag{4.180}$$

where Δ_2 is the confidence interval of the specified device, which is ± 0.01 pH.

Since the standard potential of the *AgCl* electrode is $E = 0.205$ V and is calibrated at a temperature of $t_c = 25°$C and is characterized by the existence of the coefficient bias, which is equal to $k_t = 0.73$ mV/°C, then the uncertainty that will arise when the ambient temperature deviates from the calibration temperature can be calculated by the equation:

$$u_{B3} = \frac{|t_c - t_m|}{\sqrt{3}} k_t = \frac{|25 - 20|}{1.73} 0.73 = 2.11\text{ mV}, \tag{4.181}$$

where t_m is the temperature in the laboratory, measured during experiments ($t_m = 20°$C).

To find the combined u_{Bc} uncertainty of Type B, one must go to the relative values of the uncertainties $\tilde{u}$, since they have different dimensions (u_{B1}, u_{B2}, u_{B3}), and they can be calculated using the equation:

$$\tilde{u}_{Bi} = \frac{u_{Bi}}{|x|} 100\%, \tag{4.182}$$

where u_{Bi} are Type B uncertainties, and x is the estimate of the measured (investigated) physical quantity.

The relative value of uncertainty of Type B, which occurs during the manufacture of measuring needles, associated with the deviation of the resulting radii of the heads, we calculate by the equation:

$$\tilde{u}_{BR} = \frac{u_{BR}}{|\bar{R}|} 100\% = \frac{0.75}{45.78} 100\% = 1.64\%. \tag{4.183}$$

Relative Type B uncertainty is associated with the permissible deviation of the potential from its nominal value of $E = 0.205$ V will be:

$$\tilde{u}_{B1} = \frac{u_{B1}}{|E|} 100\% = \frac{1.73 \cdot 10^{-3}}{0.205} 100\% = 0.84\%. \tag{4.184}$$

Considering the Type B uncertainty u_{B2} (4.180) associated with determining the concentration in this particular experiment, when preparing a standard *KCl* solution, the relative uncertainty will be:

$$\tilde{u}_{B2} = \frac{u_{B2}}{|pX|} 100\% = \frac{0.0058}{7} 100\% = 0.08\%. \tag{4.185}$$

The relative uncertainty $\tilde{u}_{B3}$ associated with the deviation of the ambient temperature from the calibration temperature of the electrodes is calculated considering the value of the electrode potential according to the equation:

$$\tilde{u}_{B3} = \frac{u_{B3}}{|E|} 100\% = \frac{2.11 \cdot 10^{-3}}{0.205} 100\% = 1.03\%. \tag{4.186}$$

To study ion current rectification, the *SP-300* potentiostat is used, which in accordance with the technical documentation, has a current measurement accuracy of $\tilde{u}_{B4} = 0.1\%$.

Considering the relative uncertainties obtained above (4.183)–(4.186), the combined uncertainty of Type B $\tilde{u}_{Bc}$ is determined by the equation:

$$\begin{aligned} \tilde{u}_{Bc} &= \left[\left(\frac{u_{BR}}{|\bar{R}|} 100\%\right)^2 + \left(\frac{u_{B1}}{|E|} 100\%\right)^2 + \left(\frac{u_{B2}}{|pX|} 100\%\right)^2 + \left(\frac{u_{B3}}{|E|} 100\%\right)^2 + \tilde{u}_{B4}^2\right]^{0.5} \\ &= \left(1.64^2 + 0.84^2 + 0.08^2 + 1.03^2 + 0.1^2\right)^{0.5} = 2.11\% \\ &= \left(0.003^2 + 0.42^2 + 0.51^2 + 0.1^2\right)^{0.5} = 0.67\%. \end{aligned} \tag{4.187}$$

The relative Type A uncertainties at positive and negative potentials were:

$$\tilde{u}_{A+} = \frac{u_{AI+}}{|\bar{I}_+|} 100\% = \frac{2.66}{|12.75|} 100\% = 20.86\%, \tag{4.188}$$

$$\tilde{u}_{A-} = \frac{u_{AI-}}{|\bar{I}_-|} 100\% = \frac{2.27}{|15|} 100\% = 15.13\%. \tag{4.189}$$

Considering the values of uncertainties of Type A ($\tilde{u}_{AI+} = 20.86\%$, $\tilde{u}_{AI-} = 15.13\%$) and Type B uncertainty $\tilde{u}_{Bc}$, the combined uncertainty of the research results is determined by the equation:

$$\tilde{u}_{c+} = \left[\tilde{u}_{AI+}{}^2 + \tilde{u}_{Bc}{}^2\right]^{0.5} = \left(20.86^2 + 2.11^2\right)^{0.5} = 20.97\%. \quad (4.190)$$

$$\tilde{u}_{c-} = \left[\tilde{u}_{AI-}{}^2 + \tilde{u}_{Bc}{}^2\right]^{0.5} = \left(15.13^2 + 2.11^2\right)^{0.5} = 15.28\%. \quad (4.191)$$

The relative values of the combined uncertainties of the ionic current measurement results at positive and negative potentials can be converted to absolute values using the equations:

$$u_{c+} = \frac{\left[\tilde{u}_{AI+}{}^2 + \tilde{u}_{Bc}{}^2\right]^{0.5}}{100\%}\left|\bar{I}_+\right| = \frac{\left[20.86^2 + 2.11^2\right]^{0.5}}{100\%}|12.75| = 2.67 \text{ nA}, \quad (4.192)$$

$$u_{c-} = \frac{\left[\tilde{u}_{AI-}{}^2 + \tilde{u}_{Bc}{}^2\right]^{0.5}}{100\%}\left|\bar{I}_-\right| = \frac{\left[15.13^2 + 2.11^2\right]^{0.5}}{100\%}|15| = 2.29 \text{ nA}. \quad (4.193)$$

It is possible to present the results of studies in the form of an expanded uncertainty separately at a positive and negative potential through the combined relative uncertainties in the form:

$$U_+ = k_p(\vartheta_{\text{eff}})u_{c+}, \quad (4.194)$$

$$U_- = k_p(\vartheta_{\text{eff}})u_{c-}, \quad (4.195)$$

where ϑ_{eff} is the effective degrees of freedom, which the equations will determine:

$$\vartheta_{\text{eff}} = (n-1)\frac{u^4_{c+(-)}}{u^4_{A+(-)}}. \quad (4.196)$$

The effective degrees of freedom (4.196) for positive and negative potentials:

$$\vartheta_{\text{eff}+} = (4-1)\frac{2.67^4}{2.66^4} = 3,$$

$$\vartheta_{\text{eff}-} = (4-1)\frac{2.29^4}{2.27^4} = 3.$$

According to the values of the effective degrees of freedom ϑ_{eff} from the t-distribution table (Student's table), the values of the coverage factors are selected, which are equal to $k_t(\vartheta_{\text{eff}+}) = 3.18$ and $k_t(\vartheta_{\text{eff}-}) = 3.18$.

The corresponding values of the expanded uncertainties of measuring the current strength at positive and negative potentials are determined by formulas (4.194) and (4.195):

$$U_+ = 3.18 \cdot 2.67 = 8.49 \text{ nA for } p = 0.95,$$

$$U_- = 3.18 \cdot 2.29 = 7.28 \text{ nA for } p = 0.95.$$

Therefore, the results of studies of the ion current rectification, considering the uncertainties introduced using auxiliary equipment, can be represented as:

$$I_+ = \bar{I}_+ \pm U_+ = 12.75 \pm 8.49 \text{ nA at } p = 0.95,$$

$$I_- = \bar{I}^- \pm U_- = -15 \pm 7.28 \text{ nA at } p = 0.95.$$

4.4.15 Evaluation of Measurement Uncertainty of Blood Pressure Based on Electrocardiography (ECG) and Photoplethysmography (PPG) Signals

For the experiments, we utilized the MAX86150 Evaluation System (Figure 4.6), a validated platform designed for assessing the MAX86150 integrated PPG and single-lead ECG sensor module.[14]

Using this evaluation system, two healthy subjects' fingers were measured for ECG and PPG signals. Under ideal conditions—such as stable skin contact and minimal motion—the PPG sensor provides heart rate measurements with an accuracy of $\pm$ 3 beats per minute (bpm), which corresponds to a relative accuracy of $\gamma_1 =$ 4.3% assuming a typical resting heart rate of 70 bpm. The relative accuracy of the ECG channel is estimated at $\gamma_2 = 0.08\%$, derived from a typical input-referred noise of approximately 1.2 μV and a typical resting ECG signal amplitude of approximately 1.5 mV observed in healthy individuals.

Thus, the MAX86150 Evaluation Board demonstrates relative measurement accuracy typically under 5%, making it suitable for research-grade physiological signal monitoring. Final accuracy depends on signal processing algorithms, environmental conditions, and application-specific calibration.

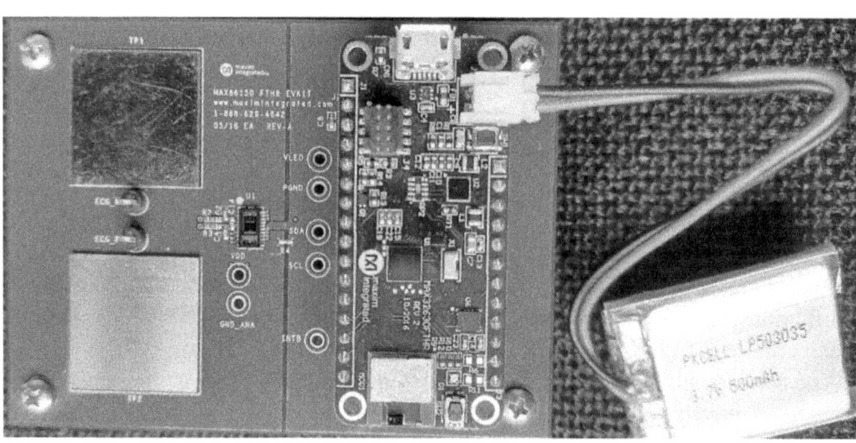

FIGURE 4.6 The MAX86150 evaluation system used for ECG and PPG signal measurements.

Based on different mathematical models for blood pressure estimation (linear and nonlinear), experimental values of systolic blood pressure (SBP) and diastolic blood pressure (DBP) were obtained for two subjects of different ages. Table 4.9 presents the results of blood pressure measurements obtained using both the linear method (Method 1) and the nonlinear method (Method 2) for the estimation of SBP and DBP. The results of blood pressure estimation using two other methods based on the determination of pressure through Mean Blood Pressure values—linear (Method 3) and nonlinear (Method 4)—are presented in Table 4.10.

The Type A evaluation of measurement uncertainty for SBP and DBP was calculated based on the results of experimental studies using the following equation:

TABLE 4.9
Experimental Data for BP Determination in Two Individuals Using Linear (Method 1) and Nonlinear (Method 2) Methods

Patient 1		Patient 2	
Method 1 (SBP/DBP, mmHg)	Method 2 (SBP/DBP, mmHg)	Method 1 (SBP/DBP, mmHg)	Method 2 (SBP/DBP, mmHg)
112.50/84.00	113.73/86.40	104.50/84.00	118.64/90.24
111.75/89.00	113.94/83.41	112.50/84.00	113.73/90.24
117.00/86.50	108.46/84.96	112.25/91.50	113.29/90.24
119.50/56.50	109.16/97.87	112.50/96.50	112.59/90.24
113.00/86.50	113.07/84.96	112.00/81.50	114.35/90.24
162.25/84.00	143.48/86.40	114.50/76.50	112.59/90.24
112.75/84.00	113.51/86.40	111.50/99.00	113.29/76.06
113.25/89.00	112.59/83.41	112.50/89.00	113.29/83.41
112.50/89.00	113.29/83.41	112.00/89.00	113.73/83.41
111.50/99.00	113.29/76.06	113.25/89.00	112.59/83.41
113.75/76.50	113.29/90.24	111.75/89.00	113.94/83.41
113.25/84.00	113.07/86.40	112.00/81.00	114.35/87.76
112.75/94.00	112.59/80.00	111.25/81.50	114.92/87.76
112.25/91.50	113.29/81.77	111.75/84.00	114.35/86.40
113.00/86.50	113.07/84.96	112.00/79.00	114.54/89.04
113.25/84.00	113.07/86.40	111.50/84.00	114.54/90.24
117.50/76.50	109.16/90.24	112.00/81.50	114.35/90.24
118.00/64.00	110.13/95.37	111.75/79.00	114.74/90.24
117.00/66.50	111.02/94.45	111.00/86.50	114.74/90.24
117.75/64.00	110.43/95.37	112.00/81.50	114.35/87.76
Mean, mmHg		Mean, mmHg	
116.70/81.75	113.68/86.92	111.73/85.35	114.15/87.54
uA, mmHg		uA, mmHg	
10.99/11.12	7.22/5.54	1.86/5.84	1.29/3.82

TABLE 4.10
Experimental Data for BP Determination in Two Individuals Using MBP, Based on the Linear SBP Estimation Method (Method 3) and the Nonlinear SBP Estimation Method (Method 4)

Patient 1		Patient 2	
Method 3 (SBP/DBP, mmHg)	Method 4 (SBP/DBP, mmHg)	Method 3 (SBP/DBP, mmHg)	Method 4 (SBP/DBP, mmHg)
112.50/83.75	112.54/83.73	112.00/84.00	113.64/83.18
111.75/84.12	114.16/82.92	111.25/84.37	115.14/82.43
110.00/85.00	117.32/81.34	111.75/84.12	114.16/82.92
113.00/83.50	111.34/84.33	112.00/84.00	113.64/82.73
112.50/81.50	112.54/83.73	112.25/83.88	113.64/83.18
113.25/83.37	110.71/84.65	112.00/84.00	113.64/83.18
112.75/83.62	111.95/84.02	112.00/84.00	113.64/83.18
113.25/83.37	110.71/84.65	112.50/83.75	112.54/83.73
111.25/84.37	115.14/82.43	111.75/84.12	114.16/82.92
112.50/83.75	112.54/83.73	110.00/85.00	117.32/81.34
111.50/84.25	114.66/82.67	112.25/83.88	113.10/83.45
113.75/83.13	109.35/85.33	112.00/84.00	113.64/83.18
115.25/82.38	104.75/87.78	111.5/84.25	114.66/82.67
113.25/83.37	110.71/84.64	113.00/83.50	111.34/84.33
112.50/83.75	112.54/83.73	112.50/83.75	112.54/83.73
112.75/83.62	111.95/84.02	113.75/83.13	109.35/85.33
112.25/83.88	113.10/83.45	112.75/83.62	111.95/84.02
113.00/83.50	111.34/84.33	113.25/83.37	110.71/84.65
113.25/83.37	110.71/84.65	112.00/84.00	113.64/83.18
112.25/83.88	113.10/83.45	113.25/83.37	110.71/84.65
114.00/83.00	108.63/85.69	111.25/84.37	115.14/82.43
114.75/82.62	106.24/86.88	112.25/83.75	112.54/83.73
117.75/81.12	121.33/79.34	111.50/84.25	114.66/82.67
Mean, mmHg		Mean, mmHg	
113.00/83.40	112.06/83.98	112.12/83.93	113.28/83.34
u_A, mmHg		u_A, mmHg	
1.52/0.86	3.39/1.71	0.79/0.40	1.72/0.87

$$u_A = \left[\frac{\sum_{i}^{n} \left(BP_i - \overline{BP} \right)^2}{n(n-1)} \right]^{0.5}. \quad (4.197)$$

The first approach to the mathematical determination of blood pressure using the linear Method 1 revealed substantially larger Type A measurement uncertainty values: for the first subject, 10.99 mmHg for SBP and 11.12 mmHg for DBP, whereas for the second subject, 1.86 mmHg for SBP and 5.84 mmHg for DBP (Table 4.9).

The Type A evaluation of measurement uncertainty obtained using the nonlinear Method 2 was 7.22 mmHg for SBP and 5.54 mmHg for DBP in the first subject, whereas in the second subject it was 1.29 mmHg for SBP and 3.82 mmHg for DBP (Table 4.9).

Using Method 3, Patient 1 was found to have a higher experimental Type A uncertainty compared to Patient 2. The maximum uncertainty values were evaluated to be 1.52 mmHg for SBP and 0.86 mmHg for DBP at the mean blood pressure of 113/83.4 mmHg. In Method 4, the maximum experimental measurement uncertainties were also observed in Patient 1, evaluated as 3.39 mmHg for SBP and 1.71 mmHg for DBP at a mean blood pressure of 112/84 mmHg (Table 4.10).

To determine the combined measurement uncertainty, which considers systematic effects arising from the limited accuracy of the measurement system, the relative Type A evaluation of uncertainty was calculated using the following equation:

$$\tilde{u}_A = \frac{u_A}{\overline{BP}} 100\%. \tag{4.198}$$

Substituting the corresponding maximum values of Type A uncertainties (Method 1) into Equation (4.198), we obtain a relative evaluation of Type A uncertainty that was 9.42% for SBP and 13.6% for DBP.

Considering the relative Type A evaluation of uncertainty and the accuracy specifications of the MCs (Type B evaluation of measurement uncertainties) as stated in the technical documentation—$\gamma_1 = 4.3\%$ for the PPG channel and $\gamma_2 = 0.08\%$ for the ECG channel—the relative combined measurement uncertainty was determined by equation:

$$\tilde{u}_c = \left(\tilde{u}_A{}^2 + \gamma_1^2 + \gamma_2^2\right)^{0.5}. \tag{4.199}$$

Thus, the relative combined measurement uncertainty of BP was 10.36% for SBP and 14.26% for DBP.

Using the coverage factor value $k_p = 1.96$ at a confidence level of $p = 0.95$, the value of the expanded uncertainty of BP measurements was calculated using the equation:

$$\mathrm{U} = \pm k_p \tilde{u}_c \overline{BP} / 100\%. \tag{4.200}$$

Substituting the obtained maximum values of the experimental uncertainty of BP measurements into Equation (4.200), it was found that the expanded uncertainty of SBP measurements was ± 23.7 mmHg, and the expanded uncertainty of DBP measurements does not exceed ± 22.85 mmHg at a confidence level of 95%.

The obtained expanded measurement uncertainty values exceed the Class A accuracy limits established by the British Hypertension Society, which require that, at a 95% confidence level, deviations in SBP and DBP measurements should not exceed ±15 mmHg. This suggests that the mathematical model implemented in Method 1 is not suitable for clinical blood pressure measurements.

Using the maximum Type A uncertainty values obtained from the blood pressure estimation models of Method 2 (Table 4.9) and recalculating the corresponding maximum relative Type A uncertainty according to Equation (4.198), the values were found to be 6.35% for SBP and 6.37% for DBP.

Then, the combined measurement uncertainty, calculated according to Equation (4.199), amounts to 7.67% for SBP and 7.69% for DBP.

Substituting the obtained relative combined measurement uncertainty values of blood pressure (Method 2) into Equation (4.200) yielded the absolute expanded measurement uncertainties, which were ±17.1 mmHg for SBP and ±13.1 mmHg for DBP. The results of the expanded measurement uncertainty evaluation indicate that Method 2 also fails to satisfy the British Hypertension Society's requirements for the permissible confidence level in SBP assessment and, consequently, cannot be recommended for clinical measurement of human blood pressure.

During the analysis of the study results (Methods 3 and 4, Table 4.10), the maximum relative Type A measurement uncertainties (Method 4) were 3.03% for SBP and 2.04% for DBP. The corresponding relative combined measurement uncertainties were 5.26% for SBP and 4.76% for DBP. The resulting expanded measurement uncertainty values, calculated according to Equation (4.200), did not exceed ±11.55 mmHg for SBP and ±7.83 mmHg for DBP.

Thus, the obtained results of expanded measurement uncertainty evaluation of blood pressure, based on the models implemented in Methods 3 and 4, comply with the Class A accuracy category defined by the British Hypertension Society, which specifies that, at a 95% confidence level, deviations in SBP and DBP measurements should not exceed ±15 mmHg. Therefore, these methods can be considered suitable for clinical blood pressure measurements.

Thus, the results of the BP studies of patient 1 can be presented as:

$$\overline{SBP} = 112.06 \pm 11.55 \text{ mmHg at } p = 95\%,$$

$$\overline{DBP} = 83.98 \pm 7.83 \text{ mmHg at } p = 95\%,$$

and for patient 2:

$$\overline{SBP} = 113.28 \pm 11.55 \text{ mmHg at } p = 95\%,$$

$$\overline{DBP} = 83.34 \pm 7.83 \text{ mmHg at } p = 95\%.$$

The model used in Method 3 exhibits even lower expanded measurement uncertainty. By substituting the obtained uncertainties into the equation for calculating expanded uncertainty (4.200), values of ±9.99 mmHg for SBP and ± 7.25 mmHg for DBP were obtained.

The corresponding measurement results for patient 1 are as follows:

$$\overline{SBP} = 113 \pm 9.99 \text{ mmHg at } p = 95\%,$$

$$\overline{DBP} = 83.4 \pm 7.25 \text{ mmHg at } p = 95\%,$$

and for patient 2:

$$\overline{SBP} = 112.12 \pm 9.99 \text{ mmHg at } p = 95\%,$$

$$\overline{DBP} = 83.93 \pm 7.25 \text{ mmHg at } p = 95\%.$$

4.4.16 Evaluation of Measurement Uncertainty of Temperature in Monitoring the Quality of the Laser Powder Bed Fusion Process in Additive Manufacturing

The primary statement of this study is to develop an experimental method for real-time, in-situ measurement of the melt pool temperature of SS 316L powder during laser melting in the EOSINT M280 system, and to establish a quantitative framework for evaluating the quality of the Laser Powder Bed Fusion (PBF-LB/M) process. This includes identifying temperature thresholds that indicate incomplete fusion or overheating and integrating these metrics into a system for automated process control and defect prevention. This research contributes to developing real-time quality control systems in additive manufacturing by enabling accurate temperature monitoring during the PBF-LB/M process. Such systems can help detect lack-of-fusion defects or overheating conditions that negatively affect the mechanical performance of manufactured parts. While demonstrated on SS 316L, the methodology can be adapted for other materials such as Inconel alloys, titanium, or aluminum powders by recalibrating the temperature measurement setup to account for material-specific emissivity and thermal properties.[15]

The MATLAB2023a software was utilized to analyze the melt pool temperature data recorded by the FLIR A700 IR camera during the PBF-LB/M process. Within the MATLAB environment, temperature arrays consisting of 90 to 100 matrices (each sized 480 × 640) were processed to identify and extract the maximum temperature values from each frame. These extracted values were then used for further processing during the creation of the object at different laser power levels (Figure 4.7).

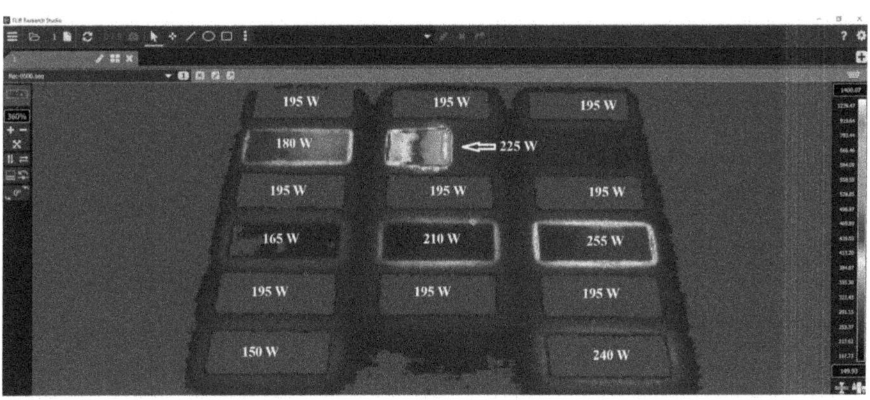

FIGURE 4.7 Image of one layer of the analyzed components during the direct laser sintering/melting of SS 316L powder at different laser power levels.

As a result of the conducted studies on the melting temperatures of SS 316L powder at various laser power levels ranging from 165 to 240 W, the average (estimated) values were obtained during the creation of 10 layers of the part (Table 4.11).

Figure 4.8 presents the experimental results of measuring the maximum melting temperatures of each track in a single layer of SS 316L powder at an LP of 195W before applying the Kalman filtering algorithm.

As indicated by the in-situ temperature measurements (see Figure 4.8) during the PBF-LB/M process, temperatures significantly higher than the melting point (liquidus temperature) are observed due to several factors:

- Optical artifacts and noise: The infrared sensor detects radiation intensity, which may not directly reflect the actual temperature due to variations in material emissivity, the presence of spatter, fumes, vapor, and reflections from the chamber walls or previous layers.
- Localized overheating and energy concentration: The laser generates a highly concentrated energy input within a small melt pool. The temperature in this area can substantially exceed the material's melting point to ensure stable melting and proper spreading of the molten metal.
- Melting and evaporation dynamics: At high laser power, localized evaporation may occur. If the sensor detects an area with significant evaporation, the measured temperature can be greatly overestimated.

TABLE 4.11
The Estimate State $\hat{T}_{k-1}$ of the Measured Temperatures during the PBF-LB/M of SS 316L Powder

	Laser Melting Power, W					
	165	180	195	210	225	240
The Number of Layers	Average Melting Temperature of Each Layer, °C					
Layer 1	1393	1426	1531	1662	1582	1609
Layer 2	1401	1451	1562	1548	1627	1608
Layer 3	1389	1477	1527	1535	1580	1622
Layer 4	1422	1426	1516	1556	1621	1619
Layer 5	1408	1487	1531	1533	1634	1623
Layer 6	1411	1460	1554	1607	1635	1643
Layer 7	1378	1477	1493	1561	1632	1632
Layer 8	1419	1477	1477	1568	1678	1637
Layer 9	1404	1487	1567	1584	1628	1631
Layer 10	1414	1487	1461	1596	1634	1628
General average temperature values	1404	1466	1522	1575	1625	1625
Standard deviation	13.2	22.7	34.0	37.3	26.5	10.6

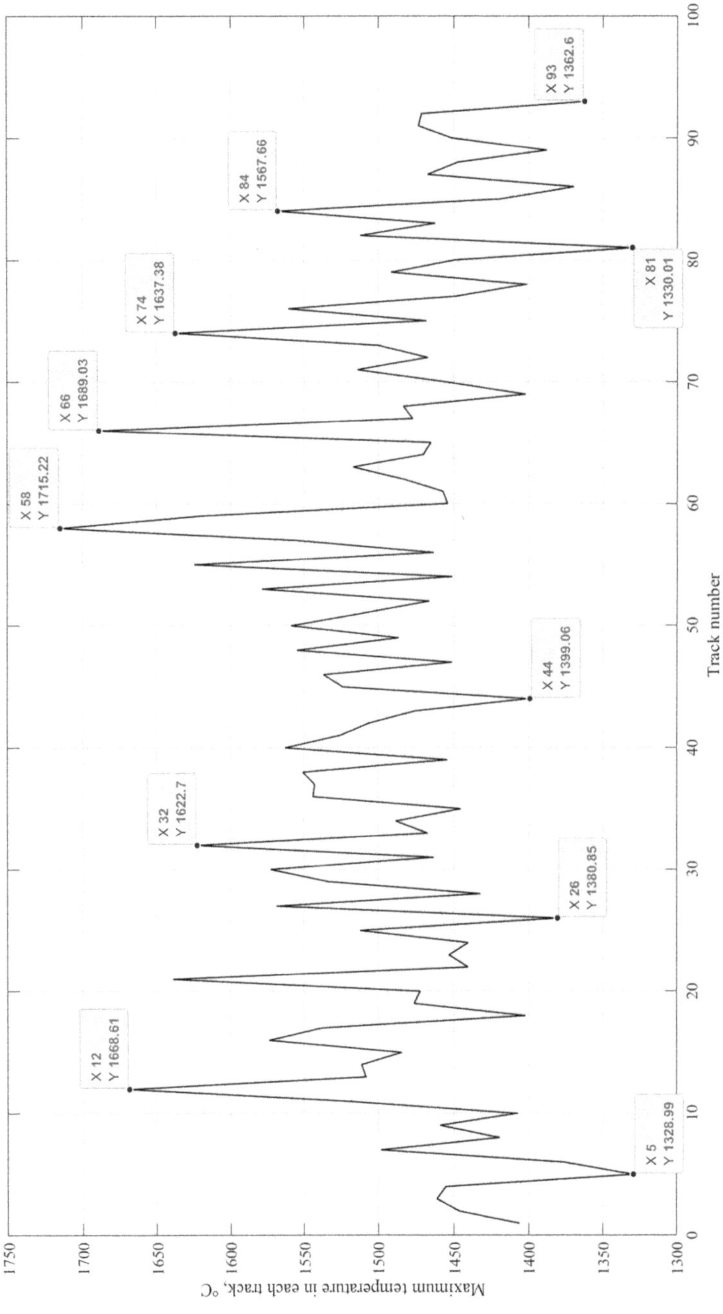

FIGURE 4.8 The results of experimental measurements of the maximum melting temperatures of a single layer (a single matrix exported to a CSV file) of SS 316L powder at an LP of 195 W before applying the Kalman filtering algorithm.

To eliminate random spikes (outliers) caused by sudden peaks that do not accurately reflect the actual melt pool temperature, and to smooth the temperature fluctuations associated with the rapid movement of the laser beam (since true temperature variations should follow a more gradual pattern), signal filtering is required. This process helps enhance the reliability of feedback signals in a closed-loop control system. Filtering prevents false commands for adjusting laser power, suppresses sharp "jumps" that are artifacts of the measurement rather than genuine process phenomena, and enables accurate determination of the metal powder's melting temperatures.

The Kalman filter is widely used in science and engineering, particularly for tracking and monitoring the state of various processes.[16,17] Experimental studies have demonstrated that the KF algorithm can be applied to the PBF-LB/M process of metal powders to estimate the melt pool temperature. Its main goal is to minimize the error caused by noise during estimation, which in turn reduces the covariance of the error under specific conditions. The KF evaluates the process states over time as a recursive function, incorporating input measurements within the process's mathematical model. The filtering algorithm works in two steps: the first step provides an estimate of the current state of the uncertain variables, while the second step updates the previous estimate using a weighted arithmetic mean after receiving a new measurement. We have presented a process model that evaluates (determines) the process states and errors as linear stochastic Equations (4.201) and (4.202):

$$\hat{T}_k^- = F\hat{T}_{k-1}, \tag{4.201}$$

$$P_k^- = FP_{k-1}F^T + Q_k, \tag{4.202}$$

where $\hat{T}_k^-$ is the predicted state of the quantity being determined (in our case, this is the predicted state of the temperature of the melt pool metal powder); F is a state transition matrix (in our case $F = 1$); $\hat{T}_{k-1}$ is the estimate of the previous state of the quantity being determined; P_k^- is the prediction error covariance; P_{k-1} is the estimate of the previous state of the prediction error covariance (variance caused by imperfections of the equipment used, which was estimated based on the combined measurement uncertainty of Type B); Q_k is the process noise matrix.

In our case, we described the process noise $Q_k = SE^2$ by the variance, which was determined based on the standard error of the mean temperature value (4.203) for a specific laser power value:

$$SE^2 = \frac{\sum_{i=1}^{n} \left(T_i - \overline{T}\right)}{n(n-1)}, \tag{4.203}$$

where T_i is the temperature values measured by the IR camera in the PBF-LB/M process at one preset laser power, n is the number of measured temperature values at the preset laser power, and $\overline{T}$ is the average temperature at the preset laser power.

The previous state estimate of the prediction error covariance, P_{k-1} was calculated by accounting for the variance arising from the limitations of the equipment used and was based on the combined uncertainty of Type B temperature measurements (4.204):

$$\tilde{u}_{Bc_{\text{inst}}} = \left[\left(\frac{u_{B\text{cam}}}{\overline{T}_{ob}} \right)^2 + \left(\frac{u_{Bta}}{|\overline{t_a}|} \right)^2 + \left(\frac{u_{Bd}}{|\overline{d}|} \right)^2 + \left(\frac{u_{B\text{hum}}}{|\overline{\gamma}|} \right)^2 \right]^{0.5}, \tag{4.204}$$

where $\overline{T}_{ob}$ represents the first estimated average value of the maximum melt pool temperature of the SS 316L powder within the LP range of 165 to 240 W, which was 1625°C (Table 4.11); $u_{B\text{cam}}$ represents the measurement uncertainty of the temperature of Type B caused by the relative error of the IR camera ($\delta = 2\%$); $u_{B\text{cam}} = \frac{\delta}{100\% \cdot \sqrt{3}} \overline{T}_{ob} = \frac{2\%}{1.73 \cdot 100\%} 1625°\text{C} = 18.8°\text{C}$; $\overline{t_a}$ is the ambient temperature inside the building chamber of the EOSINT M280 system during the experiments, which was 33.5°C; $u_{Bta} = \frac{\Delta_{ta}}{\sqrt{3}} = \frac{\pm 1.1°\text{C}}{1.73} = 0.64°\text{C}$ is the Type B ambient temperature measurement uncertainty, which is introduced by the thermocouple due to its absolute error of $\Delta_{ta} = \pm 1.1°\text{C}$, assuming a uniform distribution of the error; $\overline{d}$ is the distance from the IR camera to the objects of study, which was 0.28 m; $u_{Bd} = \frac{\Delta_d}{\sqrt{3}} = \frac{10^{-3} m}{1.73} = 0.58$ mm is the Type B distance measurement uncertainty, caused by a measurement error of $\Delta_d = \pm 1$ mm, was calculated assuming a uniform distribution of the error; $\overline{\gamma}$ is the measured value of relative humidity in the laboratory during the experiments, which was 22%; $u_{B\text{hum}} = \frac{\Delta_{\text{hum}}}{\sqrt{3}} = \frac{1\%}{1.73} = 0.58\%$ is the Type B uncertainty in humidity measurement which was caused by the existence of device error ($\Delta_{\text{hum}} = \pm 1\%$ RH).

By substituting the calculated standard uncertainties, estimated values of the measured maximum melting temperature, and additional parameters affecting the measurement process into formula (4.204), the maximum relative combined uncertainty of Type B was determined to be $\tilde{u}_{Bc_{\text{inst}}} = 0.035$ (or 3.5%). Considering that in the laser power range from 165 to 240 W, the estimated maximum average melt pool temperature was 1625°C at the laser power of 240 W, the maximum combined uncertainty of Type B, calculated based on the relative combined uncertainty $\tilde{u}_{Bc_{\text{inst}}}$, was $u_{Bc_{\text{inst}}} = \overline{T}_{ob} \times \tilde{u}_{Bc_{\text{inst}}} = 1625°\text{C} \times 0.035 = 57°\text{C}$. Therefore, the variance defined by the combined measurement uncertainty of Type B was used as the estimate for the previous state of the prediction error covariance $P_{k-1} = u^2_{Bc_{\text{inst}}} = 57^2$.

The subsequent step entails determining a posterior estimate of the melt pool temperature $\hat{T}_k$ by combining the prior estimate $\hat{T}_k^-$ (4.201) with a weighted difference between the actual temperature measurement z_k obtained from the IR camera and the predicted measurement $H\hat{T}_k^-$, as expressed in Equation (4.205):

$$\hat{T}_k = \hat{T}_k^- + K_k\left(z_k - H\hat{T}_k^-\right) = \hat{T}_k^-(1 - K_k H) + K_k z_k, \tag{4.205}$$

where H is a parameter that indicates how the measurements and the process states are related (in our case, $H = 1$); K_k is the gain factor that minimizes a posterior error $P_k = P_k^- - K_k H P_k^-$ covariance equation:

$$K_k = P_k^- H^T \left(H P_k^- H^T + R_k\right)^{-1}, \tag{4.206}$$

where R_k is the measurement error covariance matrix (the measurement noise matrix). In our case, the value of $R_k = u_A^2$ was defined as the variance of the melt pool temperature measurements, which was calculated based on the standard measurement uncertainty of Type A u_A for a specific laser power value that is described by Equation (4.207):

$$u_A^2 = \frac{\sum_{i=1}^{n} \left(T_i - \overline{T}\right)}{n-1}. \tag{4.207}$$

By utilizing Equations (4.201) through (4.207), the Kalman filter can be applied to the temperature measurement data acquired from the IR camera and stored in CSV files to estimate the melt pool temperature for each track throughout the PBF-LB/M process. Equations (4.201) through (4.207) were implemented in MATLAB to calculate the melt pool temperatures of SS 316L powder using temperature data recorded by an IR camera and saved in CSV files.

Using the temperature estimates obtained for the ten laser melting layers (Table 4.11), general average temperature values were calculated and used as the reference for the previous temperature estimates $\hat{T}_{k-1}$ (4.201) and (4.205) in the Kalman filtering process (Figure 4.9). These average previous temperature estimates (reference values for the initial temperature estimates) were determined with a maximum standard temperature deviation of $\text{StDev}_{\max} = 37.3°\text{C}$ (Table 4.11).

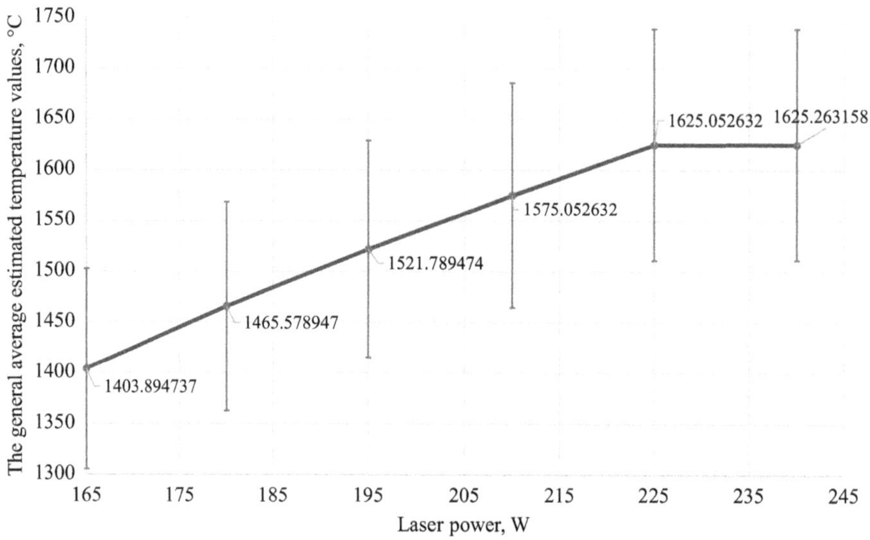

FIGURE 4.9 The average melting temperatures of all studied layers of SS 316L powder at various laser power levels.

Thus, based on the measurements of the ten repeated layers, the maximum experimental Type A uncertainty was calculated using the following equation:

$$u_{A\max}\left(\overline{T}_{ob}\right) = \frac{\text{StDev}_{\max}}{\sqrt{10}} = \frac{37.3}{3.16} = 11.8°\text{C}. \quad (4.208)$$

This maximum experimental Type A uncertainty (4.208) will be used in the application of Kalman filtering as the process noise $Q_k = u^2_{A\max}\left(\overline{T}_{ob}\right) = (11.8)$.

Considering the combined Type B uncertainty $u_{Bc_{\text{inst}}} = 57°\text{C}$ derived from formula (4.204) and the maximum experimental Type A uncertainty $u_{A\max}\left(\overline{T}_{ob}\right) = 11.8°\text{C}$ (4.208), the combined uncertainty of temperature measurements was obtained using Equation (4.209):

$$u_c = \left({u_{Bc_{\text{inst}}}}^2 + u^2_{A\max}\left(\overline{T}_{ob}\right)\right)^{0.5} = 58.2°\text{C}. \quad (4.209)$$

The expanded uncertainty U of temperature measurements during the PBF-LB/M process for SS 316L powder was determined using Equation (4.210), resulting in 114°C at a $p = 95\%$ confidence level:

$$U = \pm k_p u_c = \pm 1.96 * 58.2 = \pm 114°\text{C at } p = 95\%, \quad (4.210)$$

where k_p is the coverage coefficient for a normal distribution with probability p, the value of which is taken from Student's t-distribution table, and for probability $p = 0.95$ takes the value $k_p = 1.96$.

The expanded uncertainty of temperature measurement, expressed as a relative value $\tilde{U}$, was $\pm 7.02\%$ (4.211):

$$\tilde{U} = \frac{U}{\overline{T}_{ob_{\max}}} 100\% = \frac{114}{1625} 100\% = \pm 7.02\%. \quad (4.211)$$

In the MATLAB environment, based on the measured temperature values of each track using the FLIR A700 IR camera at laser melting power levels from 165 to 240 W and the obtained average estimated temperature values (Figure 4.9), which were used as estimates of previous states $\hat{T}_{k-1}$, filtering was applied using models (4.201)–(4.207) to the temperature measurement results extracted from CSV files. Implementing the proposed filtering method for the measured temperature values during the PBF-LB/M process at various laser power levels can determine each melt track's actual melt pool temperature.

As a result of applying the proposed filtering algorithm (4.201)–(4.207) to the experimental measurements of the maximum temperature values for each track extracted from the data arrays, the actual melt pool temperature values were determined for each track at different laser power levels. Figure 4.10 presents the obtained results for determining the melting temperatures of each track after applying Kalman filtering during the creation of layer 5 at laser sintering/melting power levels of 165, 180, 195, 210, and 225 W.

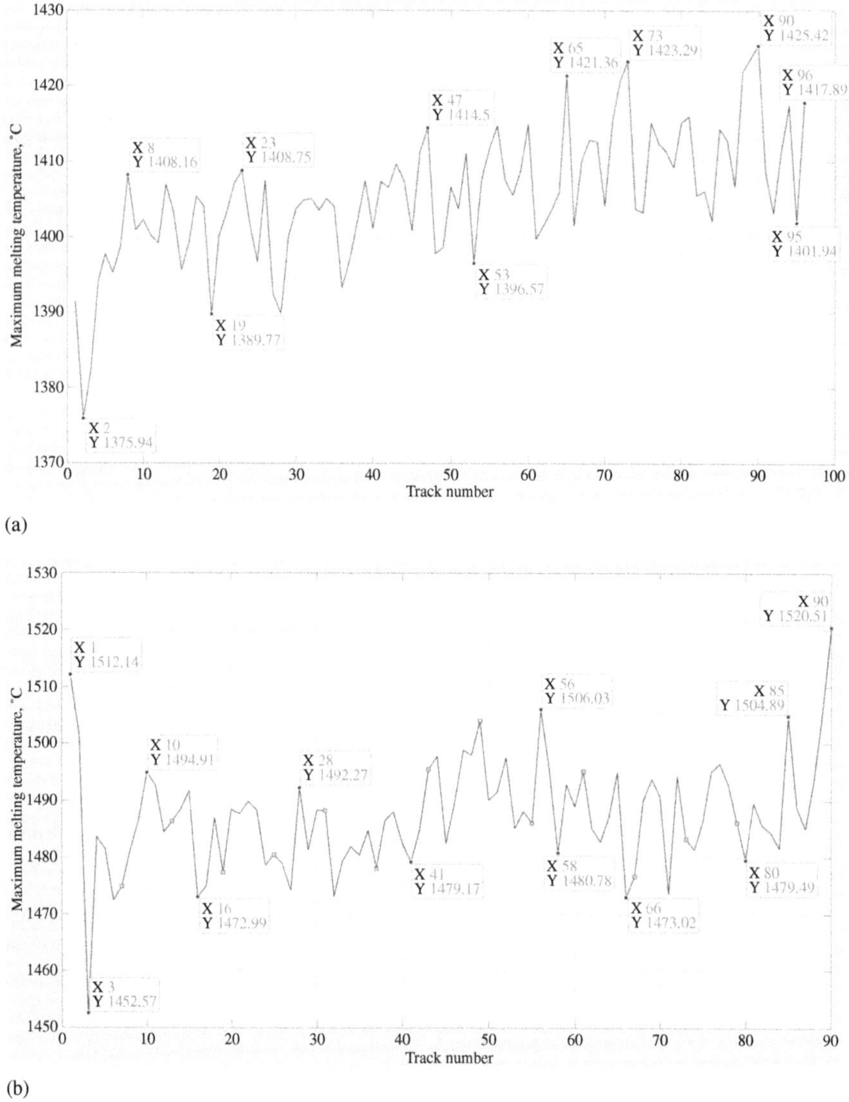

FIGURE 4.10 The sintering/melting temperature values for each track determined after applying Kalman filtering to the temperature measurement results: (a) at a laser power of 165 W. (b) at 180 W.

Thus, by using the method for determining the melt pool temperature outlined above, which is based on temperature measurements from the IR camera and the mathematical filtering models (4.201)–(4.207), the actual melt pool temperature values for each melt track can be determined.

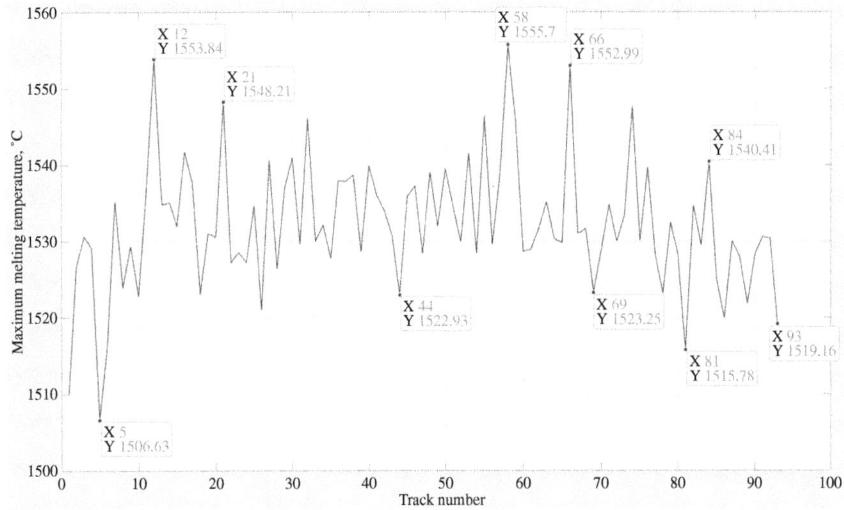

(c)

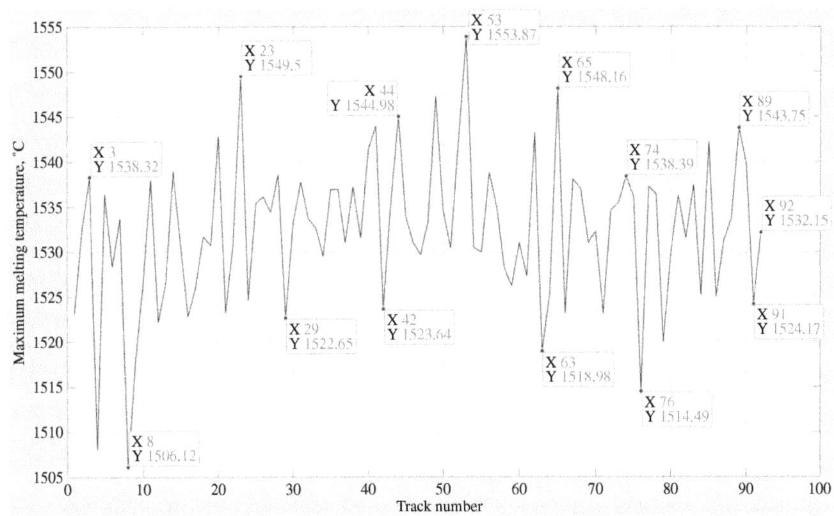

(d)

FIGURE 4.10 (Continued) (c) at 195 W. (d) at 210 W.

As shown in Table 4.11 and Figure 4.9, the study results of the average sintering/melting temperatures of SS 316L powder indicate that at the laser power of 165 W, the average sintering temperature was 1404°C. It does not reach the melting point (liquidus temperature) of SS 316L powder and, therefore, cannot melt each track during the DMLM process, leading to a lack of fusion and the formation of porous components with low density.

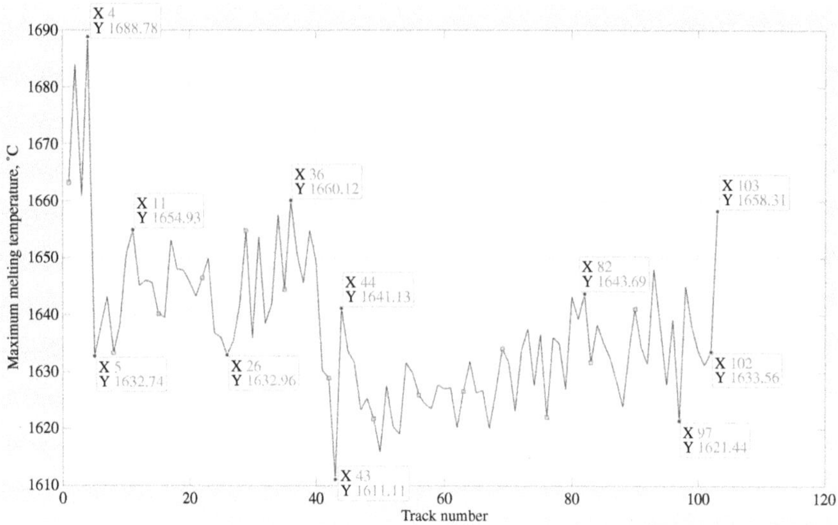

(e)

FIGURE 4.10 (Continued) (e) at 225 W.

At a laser melting power of 180 W, the average melting temperature of the SS 316L powder was 1466°C (1739 K), which corresponds to the melting point. The study results indicated that the average melting temperature of the SS 316L powder ranged from 1426°C (layers 1 and 4) to 1487°C (layers 5, 9, and 10) at a laser power of 180 W (see Table 4.11 and Figure 4.10b). Furthermore, at the recommended laser power of 195 W specified by the manufacturer of the EOSINT M280 system for melting SS 316L powder, the general average melting temperature was 1522°C. The average melting temperatures of individual 316L powder layers ranged from 1461°C (layer 10, Table 4.11) to 1567°C (layer 9, Table 4.11). The melting temperatures of SS 316L powder at laser powers ranging from 210 to 240 W significantly exceeded the melting point (see Table 4.11 and Figure 4.10).

Both insufficient and excessively high melting temperatures of metal powders can negatively impact the quality and mechanical properties of the final part. To address this, we proposed a methodology for evaluating the quality of the PBF-LB/M process, based on melt pool temperature measurements obtained using the Kalman filtering algorithm and acceptable expanded uncertainty values.

A three-level quality assessment framework—typically defined as "good," "satisfactory," and "unsatisfactory"—has become a standard method for evaluating processes and outcomes across numerous industries.

Categorizing process quality into three distinct levels provides a clear and effective way to classify manufactured parts, facilitating precise process control and enabling early detection of defects or deviations from target parameters. This approach supports timely corrective interventions, reducing the likelihood of producing defective components. Moreover, the simplicity of the three-level system enhances the efficiency of process monitoring by enabling automated classification of production states and ensuring consistent in-situ quality control.

In our approach, the process is categorized into three quality levels: good/high (A), satisfactory (B), and unsatisfactory (C).[18]

The threshold values defining these quality levels in the PBF-LB/M process were established empirically to reflect the specific conditions of laser melting applied to 316L stainless steel powder. These classification thresholds were determined through extensive in-situ temperature monitoring and rigorous statistical analysis. By analyzing temperature variation trends, incorporating measurement uncertainty, and calculating tolerance intervals at varying laser power levels, we ensured that the proposed criteria aligned with the experimentally validated stability of the PBF-LB/M process.

A good/high quality indicator (A) is represented by the measured melt pool temperature values, which have undergone Kalman filtering and fall within an acceptable deviation range relative to the expanded uncertainty of ±7.02% (see Figure 4.9).

The maximum and minimum temperature values, determined by the lower and upper bounds of the acceptable expanded uncertainty for each laser melting power level (for instance, at a laser power of 195 W, the acceptable temperature range, accounting for expanded uncertainty, is from 1415 to 1628°C), are also defined with a relative expanded uncertainty of ±7.02%. Therefore, the uncertainty ranges can also be expressed as +7.02% above the maximum temperature value and −7.02% below the minimum temperature value. We propose classifying temperature values that fall within the range from the maximum acceptable value (where quality is still considered good A) plus the expanded uncertainty (in Figure 4.11, the range between the upper lines with the inscription Quality B), and from the minimum acceptable value (where quality is still considered good A) minus the expanded uncertainty (in Figure 4.11, the range between the lower lines labeled Quality B), as satisfactory (B). If, as a result of monitoring the temperature using Kalman filtering, the melt pool temperature exceeds the allowable relative expanded uncertainty of ± 7.02%, the quality will be classified as satisfactory (B).

The temperature values obtained after Kalman filtering that surpass the threshold for satisfactory quality (B)—either exceeding the maximum acceptable temperature plus the expanded uncertainty (in Figure 4.11 this is above the topmost line labeled Quality C) or falling below the minimum acceptable temperature minus the expanded uncertainty (in Figure 4.11, this is below the lowest line labeled Quality C)—should be classified as indicating unsatisfactory/unacceptable quality (C) in the PBF-LB/M process.

To assess the quality of the PBF-LB/M process for each track in every layer (demonstrating the methodology for process quality monitoring), we extracted the maximum (dashed line with two dots) and minimum (dashed dotted line with one dot) track temperatures during the creation of 10 layers of the part, as presented in Figure 4.11.

As seen in Figure 4.11, most of the minimum and maximum sintering/melting temperatures of 316L steel powder fall within the acceptable range of expanded uncertainty, where the quality was classified as good/high (A). However, Figure 4.11

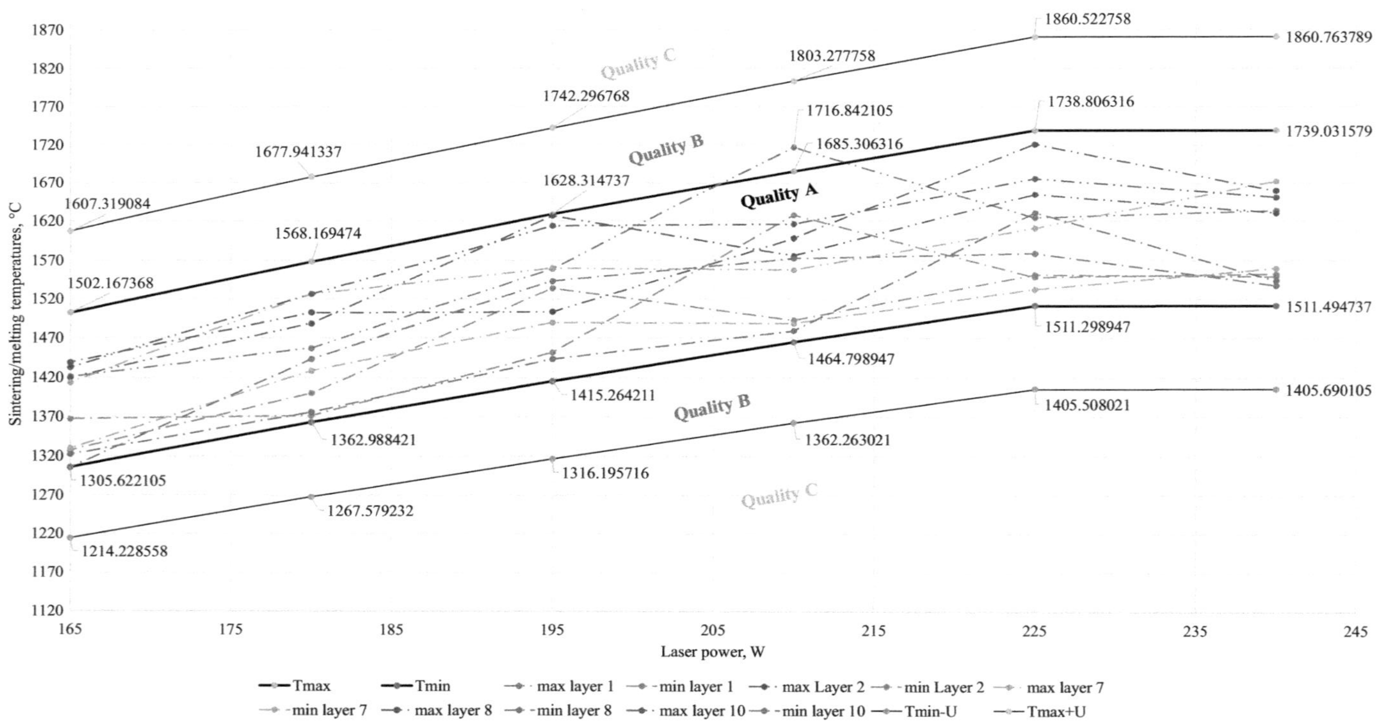

FIGURE 4.11 Maximum and minimum sintering/melting temperatures during the quality monitoring of the PBF-LB/M process of SS 316L powder.

also shows that at a laser power of 210 W during the creation of layer 1, the maximum temperature value crosses the threshold of good quality (A) and falls into the temperature range where quality was classified as satisfactory (B). Below, we present the characteristics of sintering/melting temperature variations for each track during the creation of layer 1 at laser powers of 210 W (see Figure 4.12).

As shown in Figure 4.12, during the formation of layer 1 at a laser power of 210 W, some melt tracks (e.g., tracks 32, 40, 48, 53, and others) exceeded the maximum allowable melting temperature of 1685°C, beyond which the quality of the PBF-LB/M process can no longer be classified as good (A). Temperatures exceeding 1685°C indicate that the temperature of the laser melting process goes beyond the acceptable range of expanded uncertainty, where the quality is classified as good (A), and enters the range where the quality is considered satisfactory (B). This temperature range, where the quality of the laser melting process is considered satisfactory (B), serves as an indicator for generating a feedback control signal to adjust the laser power (in this case, a signal to decrease the laser power), thereby bringing the melting temperature of 316L steel powder back into the range, where the quality is classified as good (A).

Following fabrication of the samples at different laser powers and subsequent polishing, cross-sectional analyses were performed using a scanning electron microscope (SEM). The results of the SEM examinations of the samples are presented in Figure 4.13.

Based on the obtained images (see Figure 4.13), quantitative evaluations of density and porosity were performed using the ImageJ software. The results of these evaluations are presented in Table 4.12.

Relative density values above 98% are considered acceptable according to current industrial standards.[19,20,21,22] Many studies and industry datasheets report that LPBF 316L parts with total porosity $< 1\%$ meet common industrial acceptance criteria.[23,24] In space and aerospace programs (NASA-STD-6030, MSFC-STD-3716), density is controlled through porosity limits (e.g., total porosity $\leq 0.25\%$) and pore sizes must be less than 100 μm.[25] These thresholds are equivalent to densities ≈99.75% and above.

Examination of the porosity in manufactured SS 316L samples (Figure 4.13) at different laser powers revealed that the total porosity did not exceed 1% (Table 4.12). The pore sizes were below 100 μm, and the relative density exceeded 98%, satisfying the commonly accepted industrial criteria for additive manufacturing.

It is noteworthy that at laser powers of 165 and 180 W, pores larger than 50 μm were observed with greater frequency. In contrast, at laser powers of 210 and 225 W—exceeding the manufacturer's recommended settings for the EOSINT M280 system—the pore size remained below 50 μm, and the occurrence of porosity was lower compared to laser powers below the recommended range. However, at these elevated powers, the formation of keyholes was detected, as illustrated in Figure 4.13d (images 2 and 3) and Figure 4.13e (images 2 and 3), which is indicative of extremely high melt-pool temperatures during the processing of SS 316L.

The results of microcrack analysis using SEM after the completion of laser melting of SS 316L powder correlate well with the temperature measurements and their tolerance intervals for quality assessment obtained during the laser melting process. As shown on Figure 4.11, at a laser power of 210 W, we observed melt-track

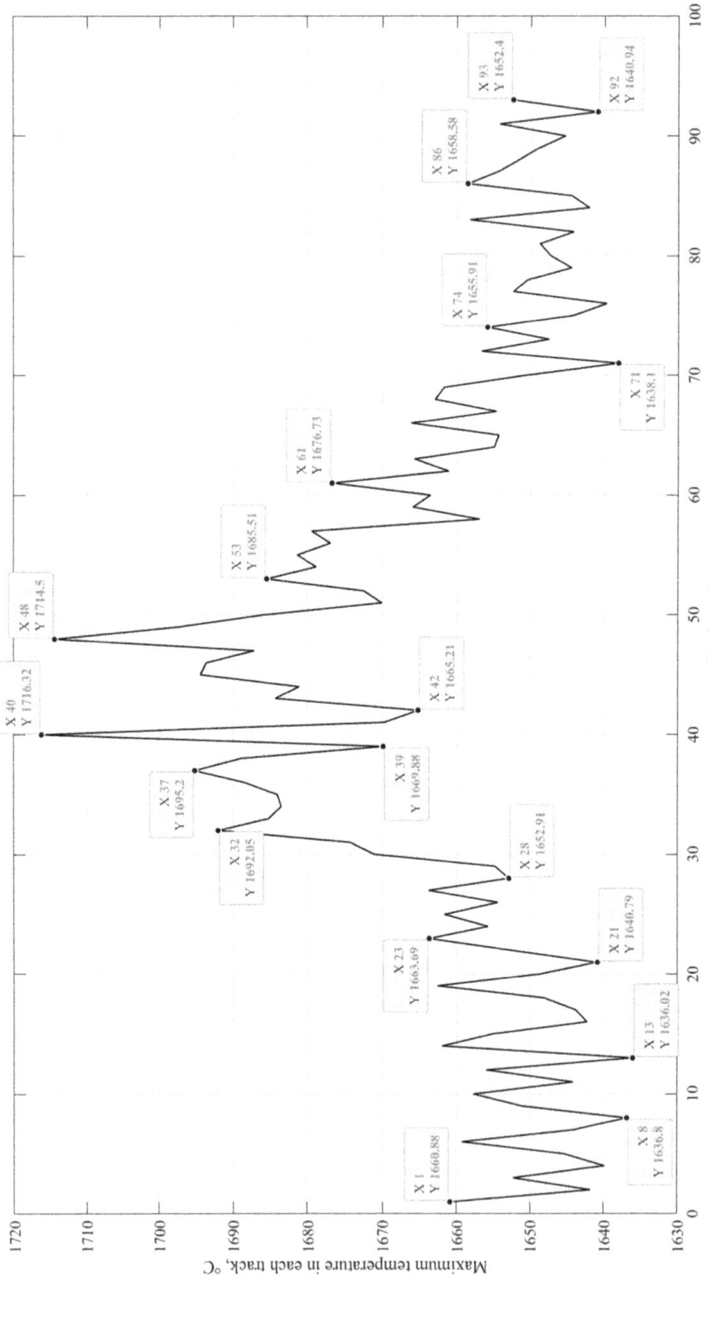

FIGURE 4.12 Temperature variation characteristics of each track's melt pool during the formation of layer 1 at a laser power of 210 W.

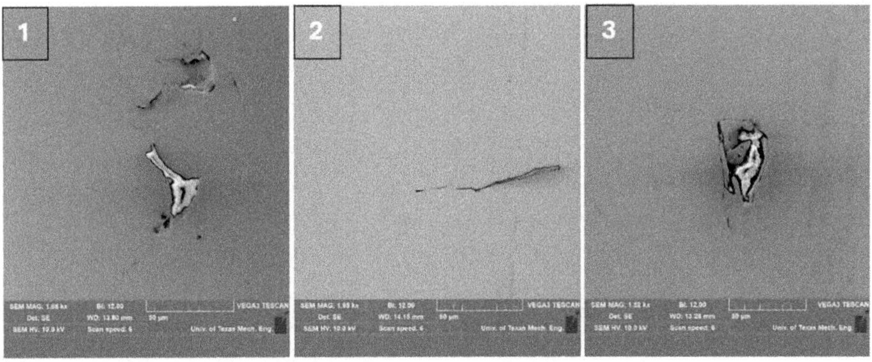

(a)

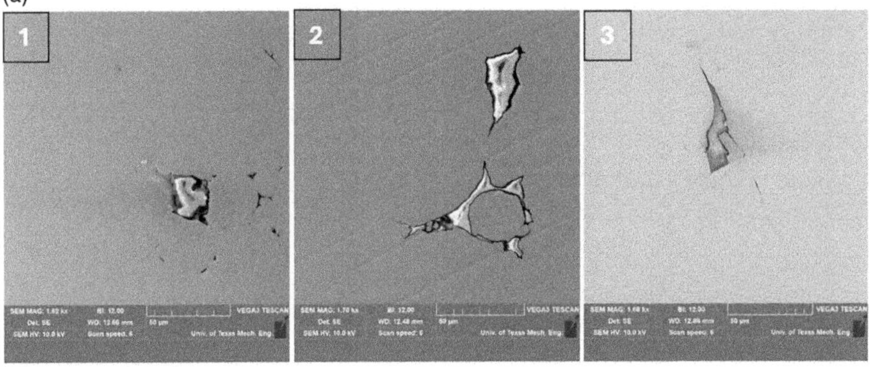

(b)

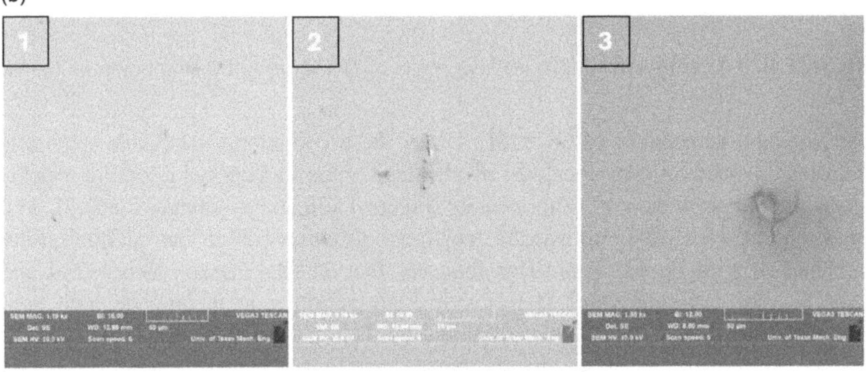

(c)

FIGURE 4.13 SEM images of large pores detected in samples fabricated at different laser powers. (a) At the laser power of 165 W. (b) At the laser power of 180 W. (c) At the laser power of 195 W.

temperatures that significantly exceeded the melting temperature at which the laser melting process quality was classified as satisfactory (B), leading to the formation of keyholes (Figure 4.13d—images 2 and 3). Experimental results demonstrate that, at laser powers exceeding the values recommended by the EOSINT M280 system manufacturer (210 and 225 W), the measured melt temperatures substantially surpass

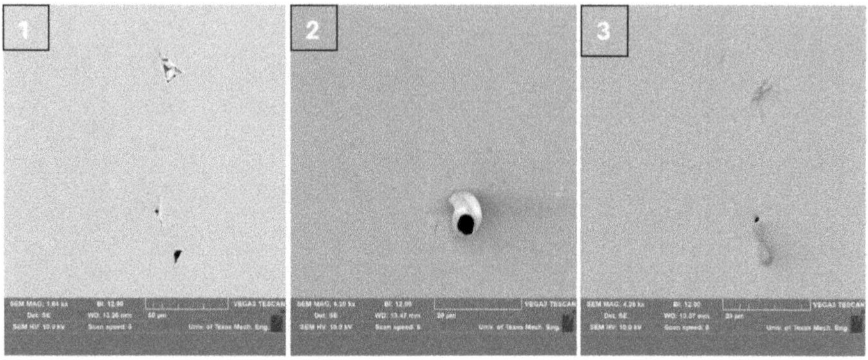

(d)

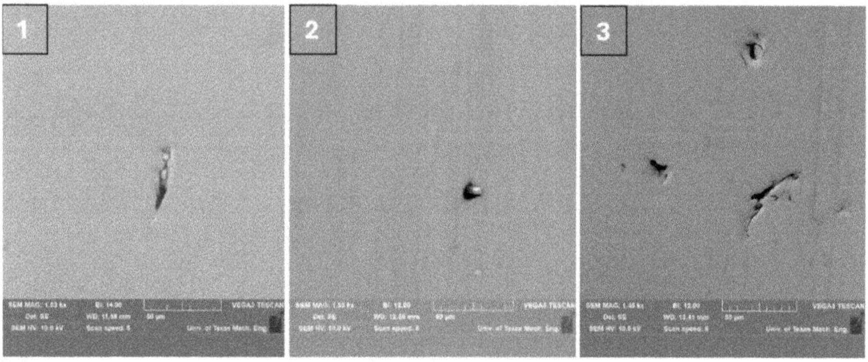

(e)

FIGURE 4.13 (Continued) (d) At the laser power of 210 W. (e) At the laser power of 225 W.

the liquidus temperature of SS 316L. Under these conditions, a keyhole regime is initiated, characterized by localized overheating of the molten metal, partial evaporation, and the formation of a vapor cavity (Figure 4.13d and e—images 2 and 3). At a scanning speed of 1083 mm/s, the melt pool solidifies before the cavity is fully replenished with liquid metal. This indicates that, at these laser powers (210 and 225 W), the scanning speed is relatively high, resulting in incomplete melt pool consolidation. Upon solidification, residual unfilled cavities serve as stress concentrators, thereby promoting premature fatigue crack initiation under cyclic loading—a highly undesirable outcome for structural applications. Nonetheless, the sizes of individual pores did not exceed 100 μm, and the measured average absolute density remained above 7.9 g/cm^3 (Table 4.12), satisfying the generally accepted industrial acceptance criteria.

Thus, as shown by the research, the proposed methodology for monitoring the quality of the laser melting process facilitates quality control of the manufactured parts during production, based on the calculated expanded uncertainty of temperature measurements. Furthermore, it enables the generation of laser power control signals if the temperature exceeds the acceptable threshold, beyond which the quality of the PBF-LB/M process of metal powders is no longer classified as good (A).

TABLE 4.12
Quantitative Evaluation of SS 316L Density and Porosity

	Laser Melting Power, W				
	165	180	195	210	225
Minimum relative density, %	98.31	98.27	99.01	99.64	98.83
Minimum absolute density, g/cm^3	7.85	7.85	7.911	7.96	7.896
Average relative density, %	98.87	99.07	99.47	99.75	99.47
Average absolute density, g/cm^3	7.899	7.92	7.95	7.97	7.947
Maximum porosity, %	1.693	1.733	0.989	0.355	1.172
Total porosity, %	0.766	0.604	0.345	0.217	0.492

The research found that at the laser power of 165 W, the average sintering temperature of 316L steel powder does not reach the melting point (Table 4.11, Figures 4.9 and 4.10a) and, therefore, cannot be recommended for use in additive manufacturing of parts using the EOSINT M280 system. Frequent porosity is observed at this laser power, which is attributed to incomplete fusion (Figure 4.13a). Although the sizes of individual pores do not exceed 100 μm and the average density is greater than 98%. At a laser power of 180 W, the average melting temperature of 316L steel powder was 1466°C. This temperature corresponds to the melting point of SS 316L powder. However, as demonstrated by the experimental results (Table 4.11), during the formation of several layers (layers 1 and 4), the average melting temperatures of 316L stainless steel powder did not reach the liquidus temperature. This indicates the necessity of using a higher laser power to ensure full melting of each track during the formation of each layer (Figure 4.13b). Therefore, the laser power of 195 W recommended by the manufacturer of the EOSINT M280 system is considered optimal. As shown by the experimental results, at this power level, the general average melting temperature of 316L powder was 1522°C with an expanded measurement uncertainty, yielding a confidence interval from 1415 to 1628°C (Figure 4.11). Within this range, the quality of the PBF-LB/M process is classified as good (A). Under these conditions, porosity occurs less frequently compared to other laser powers, with pore sizes below 50 μm (Figure 4.13c) and an average relative density of 99.5%. Research on the melting temperatures of SS 316L powder at laser powers from 210 to 240 W showed that the average melting temperature at 210 W was 1575°C, while at 225 and 240 W, the average melting temperature was 1625°C (Figure 4.9). These temperature values are much higher than the melting point of SS 316L powder and cannot be recommended for producing high-quality parts, as such temperatures lead to the formation of keyhole pores (Figure 4.13d and e), which can initiate cracks during cooling or under mechanical loads.

4.4.17 Methodology for Adjusting the Calibration Interval of Measuring Instruments Based on the Uncertainty Concept

To introduce MIs for EM rotational parameters into industrial use, conformity assessment must be carried out and an appropriate recalibration interval established. This procedure follows a structured methodology consisting of the following stages[26,27]:

1. Experimental evaluation of Type A uncertainty.
   Measurements are performed at the lower, middle, and upper points of the MI's measurement range using reference standards or calibration signals. From the resulting data, Type A standard uncertainties are calculated, and the maximum value is selected for further use.
2. Evaluation of Type B uncertainty.
   Type B standard uncertainties are assessed using available information, including prior measurement data, instrument specifications, environmental conditions, and physical properties of the measurand. A combined Type B uncertainty is then computed according to accepted formulations.
3. Determination of combined uncertainty.
   The combined measurement uncertainty is obtained using the maximum Type A standard uncertainty and the combined Type B uncertainty. In the absence of correlations, Equation (1.33) is applied; otherwise, Equation (1.43) is used.
4. Calculation of expanded uncertainty.
   The expanded uncertainty is derived by multiplying the combined uncertainty by a coverage factor, in accordance with Equation (1.48). The effective degrees of freedom are determined using the Welch–Satterthwaite formula (1.50).
5. Assessment of operating time.
   Based on prior information on the expected intensity of use and the mean time to failure, the calendar duration of MI operation is estimated.
6. Trial operation.
   The MI is subjected to extended testing under real operating conditions, simulating actual industrial use.
7. Uncertainty reassessment under field conditions.
   Following trial operation, the combined and expanded uncertainties are recalculated, accounting for real environmental and operational conditions (e.g., ambient temperature). This yields the expanded uncertainty U_E under practical use.
8. Determination of recalibration interval.
   Finally, assuming a symmetric distribution of uncertainties, the initial estimate of the recalibration interval T_1 is calculated according to equation:

$$T_1 = t \frac{\ln\left(\frac{U_E}{k_{2P-1} u(\bar{x})_{A\max}}\right)}{\ln\left(\frac{U_n}{k_P u(\bar{x})_{A\max}}\right)}, \tag{4.212}$$

where k_{2P-1} is the coverage factor corresponding to the confidence probability $2P-1$, that is, the probability value associated with the likelihood of the metrological fitness of the MI at the moment of determining its recalibration interval, t is the calendar duration of MI operation.

The coverage factor k_{2P-1} is determined from Student's t-distribution table based on the confidence probability $2P-1$ and the effective degrees of freedom ν_{eff}.

1. The second estimate of the recalibration interval of the MI T_2 is expressed by equation:

$$T_2 = t\frac{U_E - k_{2P-1}u(\bar{x})_{A\,\max}}{U_n - k_P u(\bar{x})_{A\,\max}}. \tag{4.213}$$

2. Based on the determined estimates of recalibration intervals T_1 and T_2, the recalibration interval of the MI is established as the minimum of these two values, that is:

$$T_{MI} = \min\left[T_1, T_2\right]. \tag{4.214}$$

Thus, the methodology described above for determining recalibration intervals of MIs makes it possible to establish or refine the recalibration interval on the basis of the measurement uncertainty concept. This approach complies with international requirements for the evaluation of measurement quality and contributes to ensuring the uniformity of measurements and testing.

During the validation of the methodology, a series of measurements of torque values was carried out at the lower measurement limit of 0.1 N·m, at the midpoint of the measurement range of 5 N·m, and at the upper measurement limit of 10 N·m. For this purpose, a reference torque M_{ref} was applied at the input of the MI using a set of calibrated standard weights. The equation describing the value of the reference torque takes the form:

$$M_{\text{ref}} = gRm_{\text{ref}}, \tag{4.215}$$

where g is the acceleration due to gravity, equal to 9.8066 m/s^2, R is the radius of the disk (9.9889 cm), fixed on the shaft of the EM, which is part of the torque measuring transducer, and m_{ref} is the mass of the reference weights.

To apply a torque at the lower measurement limit of the MI, a reference mass of 102.176 g must be attached to the disk of radius R, fixed on the shaft of the torque measuring transducer. The mass is suspended using a string of length l.

After setting the above parameters, a series of torque measurements is performed at the lower measurement limit and recorded in Table 4.13.

Based on the experimental data presented in Table 4.13, the Type A standard uncertainty is calculated, which for the lower measurement limit of torque is equal to:

$$u_A\left(\overline{M}_{\min}\right) = \sqrt{\frac{\sum_{i=1}^{n}\left(M_{\min,i} - \overline{M}\right)}{n(n-1)}} = 2.49 \cdot 10^{-3}\ \text{Nm}. \tag{4.216}$$

Next, the Type A standard uncertainty is determined for the reference setting at the midpoint of the torque measurement range. For this purpose, a reference mass of 5.10883 kg was attached to the disk of radius $R = 9.9889$ cm, and a series of torque measurements was performed. The results are presented in Table 4.14.

TABLE 4.13
Results of Measurements at the Lower Measurement Limit

№.	Measured Torque Values, N·m	№.	Measured Torque Values, N·m	№.	Measured Torque Values, N·m
1	0.146	8	0.138	15	0.131
2	0.117	9	0.114	16	0.125
3	0.127	10	0.115	17	0.113
4	0.121	11	0.134	18	0.114
5	0.117	12	0.117	19	0.105
6	0.132	13	0.135	20	0.106
7	0.114	14	0.126	21	0.104

TABLE 4.14
Results of Measurements at the Midpoint of the Measurement Range

№.	Measured Torque Values, N·m	№.	Measured Torque Values, N·m	№.	Measured Torque Values, N·m
1	5.057	8	5.018	15	5.037
2	5.031	9	5.055	16	5.031
3	5.049	10	5.015	17	5.025
4	5.036	11	5.058	18	5.029
5	5.055	12	5.017	19	5.035
6	5.034	13	5.025	20	5.032
7	5.029	14	5.016	21	5.049

By substituting the experimental data presented in Table 4.14, the Type A standard uncertainty (4.216) for the midpoint of the measurement range is calculated as $u_A\left(\overline{M}_{\text{midpoint}}\right) = 3.05 \cdot 10^{-3}$ N · m.

To determine the uncertainty at the upper measurement limit, a series of torque measurements was performed using a reference mass of $m_{\text{ref, max}} = 10.21765$ kg. The results of the torque measurements at the upper measurement limit are presented in Table 4.15.

By substituting the experimental data from Table 4.15, the Type A standard uncertainty (calibration uncertainty) at the upper measurement limit is calculated as $u_A\left(\overline{M}_{\max}\right) = 19.27 \cdot 10^{-3}$ N · m.

According to Step 1 of the methodology for determining the recalibration interval of the MI, the largest value of the experimental Type A standard uncertainties (calibration uncertainties) is selected from the obtained data. In this case, the maximum value corresponds to the Type A uncertainty observed at the upper measurement limit. This Type A standard uncertainty value, expressed in

TABLE 4.15
Results of Measurements at the Upper Measurement Limit

№.	Measured Torque Values, N·m	№.	Measured Torque Values, N·m	№.	Measured Torque Values, N·m
1	10.09	8	10.18	15	10.04
2	10.11	9	10.06	16	10.05
3	10.05	10	9.849	17	10.03
4	10.07	11	9.908	18	9.879
5	10.13	12	10.07	19	10.11
6	10.08	13	9.919	20	10.06
7	10.12	14	10.11	21	9.964

$u_A\left(\overline{M}_{\max}\right) = 19.27 \cdot 10^{-3}$ N·m, will be used for establishing the recalibration interval of the torque-measuring MI.

The equation describing the conversion of the EM torque MI in the static operating mode is given by:

$$M_m = Qgl = \frac{4\Omega_R S_t U_{\text{ref}}}{U_a U_b (2^q - 1)} \Delta U N g l, \tag{4.217}$$

where U_b is the supply voltage of the strain-gauge bridge sensor $(U_b = 6.5\ \text{V} \pm 0.25\ \text{mV}), \Omega_R$ is resistance of the strain-gauge bridge circuit $(\Omega_R = 3.25 \pm 0.01\ \text{k}\Omega), l$ is the length of the measuring lever connecting the tested EM shaft to the force sensor $(l = 1\ \text{m}), \Delta U$ is the output voltage of the force sensor (input voltage of the scale converter), U_a is the output voltage of the scale converter (operational amplifier), k is the gain coefficient of the operational amplifier $\left(k = Ua/\Delta U = 10^3\right), S_t$ is the sensitivity of the force sensor, U_{ref} is the reference voltage of the ADC $(U_{\text{ref}} = 3\ \text{V}), q$ is the resolution of the ADC $(q = 16)$, and N is the number of pulses at the ADC output corresponding to the measured torque value, displayed on the screen.

Next, the components of Type B standard uncertainties are evaluated, which arise due to residual systematic effects. Since the torque-measuring MI for EM rotational parameters includes a force sensor, the Type B uncertainty component introduced by the force sensor is assessed based on the existence of an uncorrected residual combined error of $\varepsilon = 0.15\%$ under the action of the maximum force $Q_{\max} = 20$ kg, according to the following expression:

$$u_{fs} = \frac{\varepsilon Q_{\max}}{\sqrt{3}100\%} = 0.15\% \frac{20}{1.73 \cdot 100\%} = 0.02\ \text{kg}. \tag{4.218}$$

Considering that the operating temperature range of the components of the torque-measuring MI for EM rotational parameters is from -10 to +40°C, the maximum possible temperature variation is $\Delta T = 50$°C. According to the specification of the scale converter (operational amplifier), which is part of the force MC, the zero-offset voltage drift of the operational amplifier is $U_{\text{drift}} = 9\,\text{nV}/°\text{C}$.

The theoretically possible (maximum) Type B standard uncertainty caused by the zero-offset voltage drift of the operational amplifier, assuming a uniform distribution, is determined by the following expression:

$$u_{B,T} = k\frac{\Delta T}{\sqrt{3}}U_{\text{drift}} = 10^3\frac{50}{1.73}9\cdot 10^{-9} = 260.12\,\mu\text{V}. \tag{4.219}$$

The Type B standard uncertainty caused by the finite resolution of the successive-approximation ADC, which accumulates pulses from the clock generator (accounting for the ADC's dead zone), with a reference voltage $U_{\text{ref}} = 3$ V and resolution $q = 16$, assuming a triangular distribution of the quantization error, is determined by the following expression:

$$u_{B,q} = \frac{LBC}{\sqrt{6}} = \frac{U_{ref}/(2^q - 1)}{\sqrt{6}} = \frac{3}{(2^{16} - 1)2.45} = 18.68\,\mu\text{V}. \tag{4.220}$$

It is also known from the specification that the allowable deviation of the supply voltage for the force sensor is $\Theta_{Ub} = \pm 0.25$ mV. The Type B standard uncertainty caused by such a deviation from the nominal voltage, assuming a uniform (rectangular) distribution, can be estimated using the following expression:

$$u_{B,\Theta u} = \frac{\Theta_{Ub}}{\sqrt{3}} = \frac{0.25}{1.73} = 0.14\,\text{mV}. \tag{4.221}$$

Since the technical documentation for the force sensor indicates that the allowable deviation of the bridge circuit resistance is $\Theta_R = \pm 10\,\Omega$, the Type B standard uncertainty arising from the residual systematic effect associated with this possible resistance deviation is calculated as:

$$u_{B,QR} = \frac{\Theta_R}{\sqrt{3}} = \frac{10}{1.73} = 5.78\,\Omega. \tag{4.222}$$

The Type B standard uncertainty caused by the residual systematic effect associated with the possible error in determining the length of the measuring lever, $\Delta l = \pm 0.01$ mm, is calculated as:

$$u_{B,l} = \frac{\Delta l}{\sqrt{3}} = \frac{0.01}{1.73} = 5.78\,\mu\text{m}. \tag{4.223}$$

The combined Type B standard uncertainty, considering the Type B standard uncertainty components calculated above, is determined as the positive square root of the combined Type B variance, which is described by the following expression:

$$u_{Bc}^2 = \left(\frac{\partial M_m}{\partial Q}\right)^2 u_{fs}{}^2 + \left(\frac{\partial M_m}{\partial U_a}\right)^2 u_{B,T}{}^2 + \left(\frac{\partial M_m}{\partial q}\right)^2 u_{B,q}{}^2 + \left(\frac{\partial M_m}{\partial U_b}\right)^2 u_{B,\Theta u}{}^2 +$$

$$+ \left(\frac{\partial M_m}{\partial \Omega_R}\right)^2 u_{B,QR}{}^2 + \left(\frac{\partial M_m}{\partial l}\right)^2 u_{B,l}{}^2, \qquad (4.224)$$

where $\frac{\partial M_m}{\partial Q} = gl = 9.81\left(\frac{m^2}{s^2}\right)$, $\frac{\partial M_m}{\partial U_a} = -\frac{4\Omega_R S_t U_{ref} gl}{kU_b U_a(2^q-1)} N = -3.33\left(\frac{\text{Nm}}{V}\right)$, $\frac{\partial M_q}{\partial q} = -\frac{4\Omega_R S_t U_{ref} gl}{kU_b(2^q-1)^2} 2^q N ln(2) = -6.93\left(\frac{\text{Nm}}{V}\right)$, $\frac{\partial M_m}{\partial U_b} = -\frac{4\Omega_R S_t U_{ref} gl}{kU_b{}^2(2^q-1)} N = -1.54\left(\frac{Nm}{V}\right)$, $\frac{\partial M_m}{\partial \Omega_R} = \frac{4S_t U_{ref} gl}{kU_b(2^q-1)} N = 3.08 \cdot 10^{-3}\left(\frac{\text{Nm}}{\Omega}\right)$, and $\frac{\partial M_m}{\partial l} = \frac{4\Omega_R S_t U_{ref} g}{kU_b(2^q-1)} N = 10(N)$ are sensitivity coefficients.

By substituting the calculated values of the sensitivity coefficients and the Type B standard uncertainties into expression (4.224), the combined Type B uncertainty is obtained as $u_{Bc} = 205.31 \cdot 10^{-3}$ N·m.

The combined uncertainty of the measurement result, considering the maximum experimental Type A uncertainty and the combined Type B uncertainty, is calculated using the following expression:

$$u_c = \sqrt{u_A\left(\overline{M}_{\max}\right)^2 + u_{Bc}{}^2} = 206.21 \cdot 10^{-3} \text{ N} \cdot \text{m}. \qquad (4.225)$$

To calculate the expanded uncertainty, which will be specified in the technical documentation for the MI, the effective number of degrees of freedom is determined using the following expression:

$$\nu_{\text{eff}} = (n-1)\frac{u_c^4}{u_A\left(\overline{M}_{\max}\right)^4} = 20\frac{206.21^4}{19.27^4} = 262266. \qquad (4.226)$$

Next, using Student's t-table, the coverage factor k_p is determined based on the effective degrees of freedom and the confidence probability, which is taken as $p = 0.95$ according to the probability information from analog measurements (confirmed by tests), resulting in $k_p = 1.96$.

Knowing the coverage factor and the combined standard uncertainty of the measurement result, the expanded uncertainty, which will be specified in the technical documentation for the investigated MI of the EM's rotational parameters, is calculated as:

$$U_n = k_p u_c = 1.96 \cdot 206.21 \cdot 10^{-3} = 0.4 \text{ N} \cdot \text{m}. \qquad (4.227)$$

According to item 5 of the methodology, assuming an operating intensity of the MI of 7 hours per day and a total usage of 3500 hours, the calendar service life t is determined to correspond to 2 calendar years.

The developed MI for the EM's rotational parameters is then put into operation. After prolonged operation, under actual operating conditions (including working temperature), a recalculation of the Type B uncertainty components and the expanded uncertainty is performed.

The Type B uncertainty due to the limited accuracy class of the force sensor remains unchanged.

The operating temperature is $T = 25°\text{C}$, and the Type B uncertainty contribution arising from the deviation of the ambient temperature from the reference conditions $T_{\text{ref}} = 20°\text{C}$, given a zero-offset voltage drift of the operational amplifier $U_{\text{drift}} = 9\ \text{nV}/°\text{C}$, is calculated as:

$$u_{B,Te} = \frac{k|T_{\text{ref}} - T|}{\sqrt{3}} U_{\text{drift}} = 10^3 \frac{|20 - 25|}{1.73} 9 \cdot 10^{-9} = 26.01\ \mu\text{V}. \tag{4.228}$$

The Type B uncertainties caused by the finite resolution of the ADC, deviations of the voltage from its nominal value, and deviations of the resistance from its nominal value remain unchanged.

By substituting the listed Type B uncertainties into Equation (4.224) for the calculation of the combined Type B uncertainty, the experimental value is obtained as $u_{BcE} = 205.29 \cdot 10^{-3}\ \text{N}\cdot\text{m}$.

The combined standard uncertainty of the measurement result under actual operating conditions is determined using the following expression:

$$u_{cE} = \sqrt{u_A\left(\overline{M}_{\max}\right)^2 + u_{Bc}{}^2} = 206.19 \cdot 10^{-3}\ \text{N}\cdot\text{m}. \tag{4.229}$$

The coverage factor k_{2P-1}, corresponding to the confidence probability $2P - 1$, i.e., the probability associated with the metrological reliability of the MI under actual operating conditions (for $p = 0.95, 2P - 1 = 2 \cdot 0.95 - 1 = 0.9$), is determined from Student's t-table using the effective degrees of freedom $\nu_{\text{eff}} > 30$ and the confidence probability $P_e = 0.9$, which gives $k_{2P-1} = 1.64$.

The expanded uncertainty U_E under actual operating conditions is given by

$$U_E = k_{2P-1} u_{cE} = 1.64 \cdot 206.19 \cdot 10^{-3} = 0.338\ \text{N}\cdot\text{m}. \tag{4.230}$$

The first estimate of the inter-calibration interval T_1 for the MI of the EM's rotational parameters is calculated using Equation (4.212), based on a calendar service life of $t = 2$ years, which corresponds to

$$T_1 = 2\frac{\ln\left(\frac{0.338}{1.64 \cdot 19.27 \cdot 10^{-3}}\right)}{\ln\left(\frac{0.4}{1.96 \cdot 19.27 \cdot 10^{-3}}\right)} = 2.01\ \text{years}.$$

The second estimate of the inter-calibration interval T_2 for the MI is calculated using Equation (4.213) and amounts to

$$T_2 = 2\frac{0.338 - 1.64 \cdot 19.27 \cdot 10^{-3}}{0.4 - 1.96 \cdot 19.27 \cdot 10^{-3}} = 1.7\ \text{years}.$$

Thus, the inter-calibration interval T_{MI} for the MI of the EM's rotational parameters is taken as the minimum of the two estimates T_1 and T_2, according to Equation (4.214), and is equal to

$$T_{MI} = \min[2.01, 1.7] = 1.7 \text{ years} = 20.4 \text{ months}.$$

It is recommended to select the inter-calibration interval of the MI in months from the series of natural numbers: [0.25; 0.5; 1; 2; 3; 4; 5; 6; 7; 8; 9; 10; 11; 12; 15; 18; 21; 24; 30], and so on in increments of 6 months.

Thus, based on the developed methodology for determining the inter-calibration interval, a metrological verification of the MI for the EM's rotational parameters was carried out. As a result, the MI was assigned standardized metrological characteristics, such as a combined standard uncertainty of measurement of $u_c = 206.21 \cdot 10^{-3}$ N·m, an expanded uncertainty of measurement of $U = 0.4$ N·m, and an inter-calibration interval of $T_{MI} = 21$ months.

4.4.18 Evaluation of Type B Measurement Uncertainty of Particle Size in Micro Cold Spray

In the measurement of particle sizes produced by micro cold spray, a FLIR Grasshopper camera (GS3-U3-51S5M-C 2/3"), a laser source (Narrowband ps lasers, NPS), a cooler (T257P-20), and a 50X Mitutoyo Plan Apo objective lens are employed. Each of these instruments is characterized by its own limited technical specifications and contributes to the combined Type B measurement uncertainty, which must be evaluated. The minimum particle size targeted for measurement is 1 μm.

We evaluate the Type B uncertainties based on the specification data of the listed equipment. According to the specification of the 50X Mitutoyo Plan Apo objective (Model Number: 378-805-3), the Numerical Aperture ($NA = 0.55$) and the Depth of Field ($DOF = \pm 0.9$) are provided. Based on the Numerical Aperture and the operating wavelength of the laser system ($\lambda = 532.15$ nm), the Rayleigh criterion can be applied to determine the theoretical lower limit for particle size measurement:

$$d_{\min} = 0.61\frac{\lambda}{NA} = 0.61\frac{0.532}{0.55} = 0.59\ \mu\text{m}. \tag{4.231}$$

Sub-pixel (localization) accuracy is the theoretical limit and is determined by the expression:

$$\sigma_{loc} = \frac{\sigma_{PSF}}{\sqrt{N_{tot}}}, \tag{4.232}$$

where N_{tot} is the total number of electrons (photoelectrons) in the Point Spread Function (PSF), $\sigma_{PSF} = \frac{d_{\min}}{2.355} = \frac{0.59}{2.355} = 0.25$ μm is SD of the Gaussian distribution.

The total number of electrons (photoelectrons) in the PSF is specified in the camera datasheet as $N_{\text{tot}} = 10208\ (e^-)$. Since the actual signal depends on illumination, exposure, laser parameters, etc., in practice a pixel often reaches only half of the full well capacity (50%), in which case we assume $N_{\text{tot}} \approx 5000$. Given this value ($N_{\text{tot}} \approx 5000$), we obtain a Type B uncertainty associated with the theoretical limit of the camera $\sigma_{\text{loc}} = \frac{0.25}{\sqrt{5000}} = 3.5$ nm.

Additionally, the camera specifications also indicate that the camera's physical pixel pitch is 3.45 μm. At 50X magnification, the effective pixel size in object space is

$$p_{\text{eff}} = \frac{3.45\mu\text{m}}{M} = \frac{3.45\mu\text{m}}{50} = 0.069\ \mu\text{m} = 69\ \text{nm/px}. \tag{4.233}$$

The Type B uncertainty due to the limited pixel pitch of the camera, assuming a rectangular distribution, is

$$u_{Bpx} = \frac{p_{\text{eff}}}{\sqrt{3}} = \frac{0.069\mu\text{m}}{1.73} = 0.04\ \mu\text{m}. \tag{4.234}$$

The combined Type B uncertainty introduced by the limited properties of the camera is

$$u_{\text{CAMcomb}} = \sqrt{u_{Bpx}^2 + \sigma_{\text{loc}}{}^2} = \sqrt{0.04^2 + 0.0035^2} = 40.2\ \text{nm}. \tag{4.235}$$

The relative combined Type B uncertainty caused by the limited characteristics of the camera, referred to the minimum particle size to be measured (1 μm), is $\tilde{u}_{\text{CAMcomb}} = \frac{0.0402}{1} 100\% = 4.02\%$.

The laser system specifications indicate that the permissible deviations for wavelength instability (λ drift/linewidth) are $\Delta\lambda = \pm 0.1$ nm. Given this value, Type B uncertainty due to wavelength instability will be equal to:

$$u(\Delta\lambda) = \frac{\Delta\lambda}{\sqrt{3}} = \frac{0.1}{1.73} = 0.058\ \text{nm}. \tag{4.236}$$

The relative Type B uncertainty at a wavelength of $\overline{\lambda} = 532.15$ nm is

$$\tilde{u}(\Delta\lambda) = \frac{u(\Delta\lambda)}{\overline{\lambda}} 100\% = \frac{0.058}{532.15} 100\% = 0.011\%. \tag{4.237}$$

Type B uncertainty caused by the instability of the laser system's output power is

$$u(P) = \frac{P_{\text{out}}}{100\%\sqrt{3}} \varepsilon = \frac{5}{100\% * 1.73} 5\% = 0.14\ \text{W}, \tag{4.238}$$

where $P_{\text{out}} = 5$ W is the output (optical) power, and $\varepsilon = 5\%$ is the permissible instability taken from the specification.

The relative $\tilde{u}(P)$ Type B uncertainty is

$$\tilde{u}(P) = \frac{u(P)}{P_{\text{out}}} 100\% = \frac{0.14}{5} 100\% = 2.8\%. \tag{4.239}$$

Type B uncertainty caused by the existence of the repetition rate tolerances ($\Delta f = \pm 1$ MHz) is

$$u(\Delta f) = \frac{\Delta f}{\sqrt{3}} = \frac{1}{1.73} = 0.58 \text{ MHz}. \tag{4.240}$$

The relative $\tilde{u}(\Delta f)$ Type B uncertainty is at a nominal repetition of $f = 40$ MHz is

$$\tilde{u}(\Delta f) = \frac{u(\Delta f)}{f} 100\% = \frac{0.58}{40} 100\% = 1.45\%. \tag{4.241}$$

Thus, the relative combined Type B uncertainty introduced by the limited properties of the laser installation is

$$\begin{aligned} u_{\text{BLcomb}} &= \sqrt{\left(\frac{u(\Delta\lambda)}{\bar{\lambda}} 100\%\right)^2 + \left(\frac{u(P)}{P_{\text{out}}} 100\%\right)^2 + \left(\frac{u(\Delta f)}{f} 100\%\right)^2} \\ &= \sqrt{\left(\frac{0.058}{532.15} 100\%\right)^2 + \left(\frac{0.14}{5} 100\%\right)^2 + \left(\frac{0.58}{40} 100\%\right)^2} \\ &= \sqrt{0.011^2 + 2.8^2 + 1.45^2} = 3.15\%. \end{aligned} \tag{4.242}$$

From the specification for the T257P-20 cooler for the laser system, it is known that the permissible deviations in temperature stability are $\Delta T = \pm 0.1$°C. Type B uncertainty caused by permissible temperature deviations is

$$u(\Delta T) = \Delta T/\sqrt{3} = 0.1/1.73 = 0.058\text{°C}. \tag{4.243}$$

If the reference temperature is maintained at $T = 20$°C, the relative Type B uncertainty will be

$$\tilde{u}(\Delta T) = \frac{u(\Delta T)}{T} 100\% = \frac{0.058}{20} 100\% = 0.29\%. \tag{4.244}$$

Considering the Depth of Field ($DOF = \pm 0.9$ μm) specified for the 50× objective lens (Model No. 378-805-3), the Type B uncertainty (Z-axis uncertainty) associated with the permissible deviation of distances along the optical axis within which the image remains sufficiently sharp (Depth of Field) is

$$u(DOF) = \frac{DOF}{\sqrt{3}} = \frac{0.9}{1.73} = 0.52 \ \mu\text{m}. \tag{4.245}$$

When measuring particles in the lower range of $PS = 1$ μm, the relative uncertainty will be

$$\tilde{u}(DOF) = \frac{u(DOF)}{PS} 100\% = \frac{0.52}{1} 100\% = 52\%. \tag{4.246}$$

The measured quantity is the particle diameter d. The Z-shift yields a change in the observed diameter Δd. To convert from position uncertainty (Z-shift) to diameter uncertainty, the laws of uncertainty propagation must be applied:

$$u(d) = \left|\frac{\partial d}{\partial z}\right| u(DOF), \tag{4.247}$$

that is, the standard uncertainty of the diameter is equal to the standard uncertainty in Z ($u(DOF)$), multiplied by the sensitivity $k = |\frac{\partial d}{\partial z}|$ (change in diameter per unit displacement in Z).

How to find the sensitivity k? We need to experimentally perform a Z-scan (focus stack): take a series of images across the Z range (e.g., ±1 μm) in 0.1 μm increments and measure the apparent diameter $d(z)$ for each position. Plot d vs z and find the slope near the working focus—this is derivative $\partial d/\partial z$. For this purpose, reference samples of particles with a size of 1 micron need to be used.

Let's assume different sensitivity values $k = |\frac{\partial d}{\partial z}|$:

If $k = 0.1$ μm/μm (the diameter changes by 0.1 μm for a 1 μm displacement), then $(d) = 0.1u(DOF) = 0.1 \times 0.52 = 0.052$ μm (5.2% of 1 μm).
If $k = 0.2$ μm/μm, then $u(d) = 0.2 \times 0.52 = 0.104$ μm (10.4%).
If $k = 0.5$ μm/μm (high sensitivity), then $u(d) = 0.5 \times 0.52 = 0.26$ μm (26%).

The explanation of the influence of the permissible Depth of Field deviation on the uncertainty of measuring the minimum particle size at a sensitivity coefficient of 0.1 is presented in Figure 4.14, and for different values of the sensitivity coefficients in Figure 4.15.

As can be seen from the results of the uncertainty analysis, the accuracy of microparticle measurements strongly depends on the sensitivity coefficient and the Type B uncertainty associated with the DOF parameter of the objective lens. This sensitivity coefficient must be determined experimentally and is highly influenced by vibrations that occur during micro cold deposition. Therefore, all equipment should be placed on optical tables to minimize the impact of vibrations on the measurement results.

Considering the relative Type B uncertainties of all equipment used, the combined uncertainty is

$$\tilde{u}_{CB} = \sqrt{\left(\left|\frac{\partial d}{\partial z}\right| \frac{u(DOF)}{PS}\right)^2 + + \tilde{u}(\Delta T)^2 + \tilde{u}_{\mathrm{CAMcomb}}{}^2 + u_{\mathrm{BLcomb}}{}^2}. \tag{4.248}$$

Thus, by substituting different values of sensitivity coefficients and Type B uncertainties calculated for each piece of equipment into Equation (4.248), we obtain the following values of the relative combined Type B uncertainty:

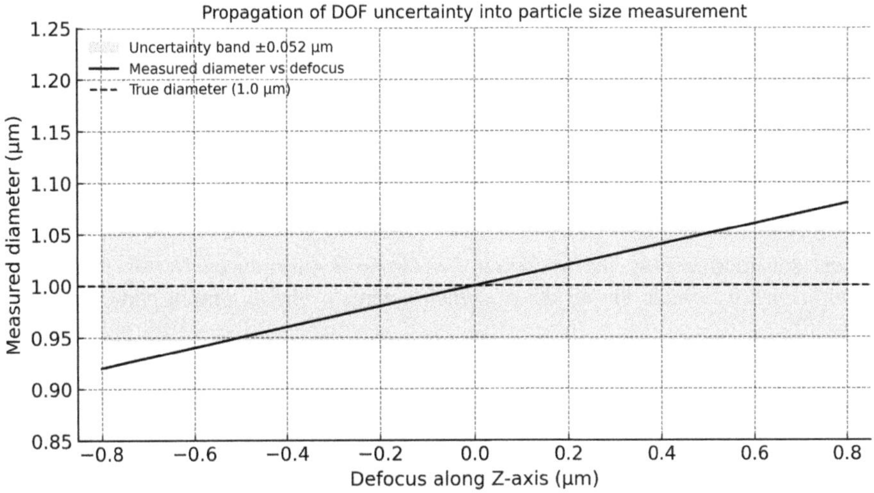

FIGURE 4.14 Influence of depth of field on particle size measurements.

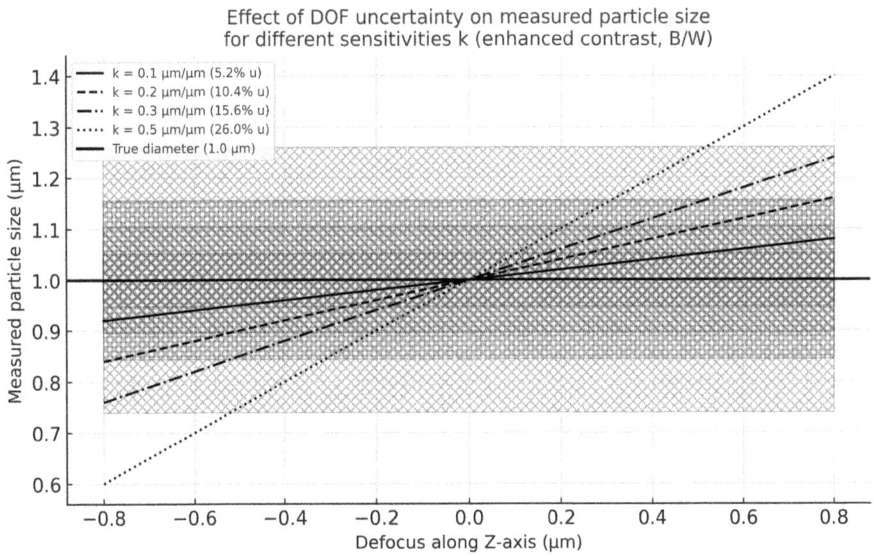

FIGURE 4.15 Influence of depth of field on particle size measurements at different sensitivity coefficients.

If $k = |\frac{\partial d}{\partial z}| = 0.1$, $\tilde{u}_{CB} = \sqrt{5.2^2 + 0.29^2 + 3.35^2 + 3.15^2} = 6.95\%$, $u_{CB} = \frac{\tilde{u}_{CB}}{100\%} d = \frac{6.95}{100\%} 1 = 0.07$ µm.

If $k = |\frac{\partial d}{\partial z}| = 0.2$, $\tilde{u}_{CB} = 11.37\%$, $u_{CB} = 0.11$ µm.

If $k = |\frac{\partial d}{\partial z}| = 0.3$, $\tilde{u}_{CB} = 16.27\%$, $u_{CB} = 0.16$ µm.

If $k = |\frac{\partial d}{\partial z}| = 0.5$, $\tilde{u}_{CB} = 26.41\%$, $u_{CB} = 0.26$ µm.

Taking the confidence level equal to 95%, we obtain an expanded uncertainty of Type B, which is

If $k = 0.1, U_B = 1.96\ u_{CB} = \pm 0.14$ μm.
If $k = 0.2, U_B = \pm 0.22$ μm.
If $k = 0.3, U_B = \pm 0.31$ μm.
If $k = 0.5, U_B = \pm 0.51$ μm.

Thus, considering only the uncertainty of Type B (introduced by the measuring equipment), we obtain the results of measurements of the minimum particle size of

$$PS = 1.0 \pm U_B = 1.0 \pm 0.14\ \mu\text{m at } k = 0.1.$$

$$PS = 1.0 \pm 0.22\ \mu\text{m at } k = 0.2.$$

$$PS = 1.0 \pm 0.31\ \mu\text{m at } k = 0.3.$$

$$PS = 1.0 \pm 0.51\ \mu\text{m at } k = 0.5.$$

4.5 RECOMMENDATIONS FOR REPORTING MEASUREMENT UNCERTAINTY

In general, when preparing a report within the measurement hierarchy, increasingly detailed information is required regarding how the measurement result, and its associated uncertainty were obtained. However, at any level of this hierarchy—covering commercial and regulatory activities in the marketplace, engineering work in industry, lower-level calibration services, industrial research and development, academic research, industrial primary standards and calibration laboratories, national measurement institutes, and the International Bureau of Weights and Measures (BIPM)—all information necessary for the re-evaluation of a measurement must be available to those who may need it.

The essential difference lies in the fact that, at the lower levels of the hierarchy, most of the required information can be made available in the form of published calibration and test reports, test specifications, calibration and test certificates, operating manuals, international and national standards, and local regulations.

When details of the measurement, including how the uncertainty of its result was evaluated, are provided by reference to existing documents—as is often the case when calibration results are reported in a certificate—it is essential that these documents are kept up to date so that they correspond directly to the measurement procedure currently in use.

A large number of measurements are carried out daily in industry and commerce without detailed reports on uncertainty. Nevertheless, most of these measurements are performed using MIs that are subject to periodic calibration or legal metrological verification. If it is known that the MIs conform to their specifications or to applicable regulatory documents, the uncertainties associated with their indications may be taken from those specifications or documents.

When preparing a report on the measurement result and its uncertainty, it is preferable to provide too much information rather than too little. Therefore, in reporting uncertainties, it is necessary to:

1. Clearly describe the methods used to compute the measurement result and its uncertainty from experimental observations and input data.
2. List all components of uncertainty and explain in detail how they were evaluated.
3. Present the data analysis in such a way that all essential steps can be easily followed and, if necessary, the reported results can be independently reproduced.
4. Provide all correction values, constants, and methods of their determination used in the analysis.

To verify whether the above requirements have been met, one should ask: '*Have I provided sufficient information, in a sufficiently clear manner, so that my result could be improved in the future if new information or new data become available?*'

The notation with the symbol "±" should, where possible, be avoided, since it is traditionally used to indicate an interval corresponding to a high level of confidence and may therefore be confused with the expanded uncertainty.

When reporting a measurement result where the measure of uncertainty is the expanded uncertainty $U = k_p \cdot u_c(y)$, the following should be provided:

1. A complete description of how the measurand Y is defined.
2. The measurement result expressed as $Y = y \pm U$, with units given for both y and U.
3. The relative expanded uncertainty $U/|y|$, $|y| \neq 0$, whenever possible.
4. The value of the coverage factor k_p used to obtain U (or both k_p and $u_c(y)$, for the convenience of those using the result).
5. An approximate confidence level associated with the interval $y \pm U$, and a statement of how it was determined.

If a measurement procedure provides more than one measurand simultaneously (i.e., it yields values for two or more output estimates y_i), then in addition to y_i and $u_c(y_i)$ for each estimate, one should report the elements of the covariance matrix $u(y_i, y_j)$ or the elements of the correlation coefficient matrix $r(y_i, y_j)$, and preferably both.

Numerical values of the estimate y and its associated standard uncertainty $u_c(y)$ or expanded uncertainty U should not be reported with an excessive number of digits. Typically, two significant digits are sufficient for $u_c(y)$ and U. In some cases, however, additional digits may need to be retained to avoid rounding errors in subsequent calculations.

When reporting final results, it is sometimes appropriate to round uncertainties upwards rather than to the nearest digit. For example, $u_c(y) = 10.47\ \mu\Omega$ may be rounded to $11\ \mu\Omega$. However, common sense should prevail, and a value such as $u(x_i) = 28.05$ kHz should be rounded to 28 kHz. Output and input estimates should be rounded consistently with their associated uncertainties. For instance, if $y = 10.05762\ \Omega$ with an uncertainty of $u(y) = 21\ \text{m}\Omega$, then the numerical value of y should be rounded to $10.058\ \Omega$. Correlation coefficients should be given to three decimal places if their absolute values are close to unity.

When providing a detailed account of how the measurement result and its uncertainty were obtained, it is necessary to follow established recommendations and, therefore:

1. give the value of each input estimate x_i together with its standard uncertainty $u(x_i)$, including a description of how they were obtained;
2. provide the evaluated covariances and correlation coefficients (preferably both) for all correlated input estimates, along with the methods used to obtain them;
3. state the degrees of freedom associated with the standard uncertainty of each input estimate and describe how they were determined;
4. present the functional relationship $Y = f(X_1, X_2, \ldots, X_N)$ and, if needed, the sensitivity coefficients.

Since the functional relationship f may be highly complex, unavailable in explicit form, or implemented only as a computer program, it is not always feasible to provide f and its derivatives. In such cases, f may be described in general terms or referenced to the program used. It is essential, however, that it be clear how the estimate of the measurand Y and its combined uncertainty $u_c(y)$ were obtained.

Notes

1 B. N. Taylor and C. E. Kuyatt, *Guidelines for evaluating and expressing the uncertainty of NIST measurement results*, NIST Technical Note 1297, Rev. ed., National Institute of Standards and Technology, Gaithersburg, MD, 2008.

2 JCGM 100:2008, *Evaluation of measurement data—Guide to the expression of uncertainty in measurement* (Int. Organ. Stand., Geneva, ISBN 50, 2008: 134).

3 O. M. Vasilevskyi, P.I. Kulakov, Elements of the theory of accuracy improvement in measurement and synchronization of angular velocities of interconnected electric motor rotors: [Monograph], 2011: 170.

4 O. M. Vasilevskyi, V. Yu. Kucheruk, Ye. T. Volodarskyi, *Uncertainty of measurement, control and test results* (OLDI-PLUS, 2020: 352).

5 ISO/IEC Guide 98-3:2008, *Uncertainty of measurement Part 3: Guide to the expression of uncertainty in measurement (GUM:1995)*, (Int. Organ. Stand., Geneva, 2008: 120).

6 O. Vasilevskyi, O. Voznyak, V. Didych, V. Sevastianov, O. Ruchka, V. Rykun, Methods for constructing high-precision potentiometric measuring instruments of ion activity, *2022 IEEE 41st International Conference on Electronics and Nanotechnology (ELNANO)*, Ukraine, 2022, pp. 247–252.

7 O. Vasilevskyi, V. Didych, O. Zabula, V. Sarana and E. M. Popovici, Developing and using measurement models to assess accuracy: using the example of measurements of the activity of ions, *2023 IEEE International Conference on Industry 4.0, Artificial Intelligence, and Communications Technology (IAICT)*, Indonesia, 2023, pp. 125–129.

8 O. M. Vasilevskyi, A. V. Podzharenko, Evaluation of the uncertainty of rotor moment of inertia measurement by the amplitude of turning vibrations, Visnyk of Vinnytsia Polytechnical Institute, 4, 2010, pp. 5–9.
9 M. Cullinan, O. Vasilevskyi, J. Allison, Methodology for determination of the emissivity of metal powders and uncertainty quantification using an infrared camera and thermocouples, *Measurement Science and Technology*, vol. 36, no. 2, p. 025013, 2025.
10 G. Mohr, S. Nowakowski, S. Altenburg, C. Maierhofer, K. Hilgenberg, Experimental Determination of the Emissivity of powder layers and bulk material in laser powder bed fusion using infrared thermography and thermocouples, *Metals*, 2020, 10(11): 1546.
11 M. Vollmer, Infrared thermal imaging. In Computer Vision: A Reference Guide, 2021, pp. 666–670. Cham: Springer International Publishing.
12 ISO 13528:2022, Statistical Methods for use in proficiency testing by interlaboratory comparison, https://www.iso.org/standard/78879.html.
13 L. Bermanec, D. Zvizdic, Interlaboratory comparison in the pressure range from 0 to 2 MPa for accredited calibration laboratories, *International Journal of Metrology and Quality Engineering*, 2015, 6(3), 307.
14 O. Vasilevskyi, R. Trishch, V. Sarana, M. Yakovlev, and E. Popovici, Experimental analysis of blood pressure estimation using electrocardiography and photoplethysmography signals from fingertip measurements, Acta IMEKO, vol. 14, no. 2, pp. 1–9, 2025.
15 O. Vasilevskyi, A. Woods, M. Jones, M. Cullinan, *In situ* monitoring and quality assessment of laser powder bed fusion process for 316L stainless steel, *The International Journal of Advanced Manufacturing Technology*, 2025, https://doi.org/10.1007/s00170-025-16448-0.
16 Kim, Y., Bang, H. *Introduction to Kalman filter and its applications: Book*, 2019, 128 p.
17 O. Vasilevskyi, A. Woods, M. Cullinan, *In Situ* monitoring and quality evaluation of the powder bed fusion process of IN718 alloy powder, *IEEE Transactions on Instrumentation and Measurement*, vol. 74, pp. 1–15, 2025.
18 O. Vasilevskyi, A. Woods, M. Jones, M. Cullinan, Quantitative methodology for assessing the quality of direct laser processing of 316L steel powder using Type I and Type II control errors. *Electronics*, 14(7), 2025, 1476.
19 Zhu, P., He, X., Guan, H., Zhang, Z., Zhang, T, Qu X. Investigation on the attainment of high-density 316L stainless steel with selective laser sintering, *Materials (Basel)*, 7(1): 110, 2023.
20 Zhang, P., Tan, J., Tian, Y. *et al.* Research progress on selective laser melting (SLM) of bulk metallic glasses (BMGs): a review, *International Journal of Advanced Manufacturing Technology*, 118, 2017–2057, 2022.
21 Bruce, D., Paradise, P., Saxena, A., Temes, S., Clark, R., Noe, C., Benedict, M., Broderick, T., Bhate, D. A critical assessment of the Archimedes density method for thin-wall specimens in laser powder bed fusion: Measurement capability, process sensitivity and property correlation, *Journal of Manufacturing Processes*, Volume 79, pp. 185–192, 2022.
22 Shanmuganathan, P.K., Purushothaman, D.B., Ponnusamy, M. Effect of High Laser Energy Density on Selective Laser Melted 316L Stainless Steel: Analysis on Metallurgical and Mechanical Properties and Comparison with Wrought 316L Stainless Steel, 3D Print Addit Manuf., 10(3): 383-392, 2023.
23 Chen-Nan Sun, Beng Loon Aw, Hengfeng Gu, Danny Ming Tak Choi, Chong Teng, Sharon Mui Ling Nai, Aravind Vasanthakumar, Chengcheng Wang, Porosity distribution

of 316 L stainless steel in laser powder bed fusion additive manufacturing due to spatial variation, Journal of Manufacturing Processes, Volume 139, Pages 81-89, 2025.
24 PNNL (Pacific Northwest National Laboratory) 36408—Studies on Printability Methodologies and Directed-Energy Deposition-Fabricated Iron Alloys for Nuclear Applications, 2024.
25 NASA-STD-6030 - Additive manufacturing requirements for spaceflight systems, 2021, https://standards.nasa.gov/sites/default/files/standards/NASA/Baseline/0/2021-04-21_nasa-std-6030-approveddocx.pdf?utm_source=chatgpt.com.
26 O. M. Vasilevskyi, Calibration method to assess the accuracy of measurement devices using the theory of uncertainty, International Journal of Metrology and Quality Engineering, 5.04: 403, 2014.
27 O. M. Vasilevskyi, Methods of determining the recalibration interval measurement tools based on the concept of uncertainty, Technical Electrodynamics, 6, 2014, pp. 81–88.

BIBLIOGRAPHY

Bermanec, L. and D. Zvizdic, Interlaboratory comparison in the pressure range from 0 to 2 MPa for accredited calibration laboratories. *International Journal of Metrology and Quality Engineering*, 6(3), (2015), 307, https://doi.org/10.1051/ijmqe/2015021

Bruce, D., P. Paradise, A. Saxena, S. Temes, R. Clark, C. Noe, M. Benedict, T. Broderick, and D. Bhate, A critical assessment of the Archimedes density method for thin-wall specimens in laser powder bed fusion: Measurement capability, process sensitivity and property correlation, *Journal of Manufacturing Processes*, 79, 2022, 185–192. https://doi.org/10.1016/j.jmapro.2022.04.059

Cullinan, M., O. Vasilevskyi, and J. Allison, Methodology for determination of the emissivity of metal powders and uncertainty quantification using an infrared camera and thermocouples, *Measurement Science and Technology*, 36(2), 2025, 025013, https://doi.org/10.1088/1361-6501/ada4c6

ISO 13528:2022, *Statistical Methods for use in proficiency testing by interlaboratory comparison*, Int. Organ. Stand., Geneva (2022). https://www.iso.org/standard/78879.html

ISO/IEC Guide 98-3:2008, *Uncertainty of measurement Part 3: Guide to the expression of uncertainty in measurement* (*GUM:1995*), Int. Organ. Stand., Geneva (2008): 120.

JCGM 100:2008, *Evaluation of measurement data—Guide to the expression of uncertainty in measurement*, Int. Organ. Stand., Geneva, ISBN 50 (2008): 134.

Kim, Y. and H. Bang, *Introduction to Kalman filter and its applications*: Book (2019), 128 p., https://books.google.com/books?hl=en&lr=&id=YhT8DwAAQBAJ&oi=fnd&pg=PA7&dq=the+Kalman+filtering+algorithm&ots=FtzATjYdO4&sig=HTfY5Ck5QwkeQPoSuG0SyuwUetM#v=onepage&q=the%20Kalman%20filtering%20algorithm&f=false

Mohr, G., S. Nowakowski, S. Altenburg, C. Maierhofer, K. Hilgenberg, Experimental determination of the emissivity of powder layers and bulk material in laser powder bed fusion using infrared thermography and thermocouples. *Metals*, 10(11), 2020, 1546. https://doi.org/10.3390/met10111546

NASA-STD-6030 – *Additive manufacturing requirements for spaceflight systems*, 2021. https://standards.nasa.gov/sites/default/files/standards/NASA/Baseline/0/2021-04-21_nasa-std-6030-approveddocx.pdf?utm_source=chatgpt.com

PNNL (Pacific Northwest National Laboratory) 36408—*Studies on printability methodologies and directed-energy deposition-fabricated iron alloys for nuclear applications*, 2024. https://www.pnnl.gov/main/publications/external/technical_reports/PNNL-36408.pdf?utm_source=chatgpt.com

Shanmuganathan, P. K., D. B. Purushothaman, and M. Ponnusamy, Effect of high laser energy density on selective laser melted 316L stainless steel: Analysis on metallurgical and mechanical properties and comparison with wrought 316L stainless steel, *3D Printing and Additive Manufacturing*, 10(3), 2023, 383–392. https://doi.org/10.1089/3dp.2021.0061

Sun, Chen-Nan, Beng Loon Aw, Hengfeng Gu, Danny Ming Tak Choi, Chong Teng, Sharon Mui Ling Nai, Aravind Vasanthakumar, and Chengcheng Wang, Porosity distribution of 316 L stainless steel in laser powder bed fusion additive manufacturing due to spatial variation, *Journal of Manufacturing Processes*, 139, 2025, 81–89. https://doi.org/10.1016/j.jmapro.2025.02.031

Taylor, B. N. and C. E. Kuyatt, *Guidelines for evaluating and expressing the uncertainty of NIST measurement results*, NIST Technical Note 1297, Rev. ed., National Institute of Standards and Technology, Gaithersburg, MD (2008). https://doi.org/10.6028/NIST.TN.1297

Vasilevskyi, O., V. Didych, O. Zabula, V. Sarana, and E. M. Popovici, Developing and using measurement models to assess accuracy: using the example of measurements of the activity of ions, *2023 IEEE International Conference on Industry 4.0, Artificial Intelligence, and Communications Technology (IAICT)*, Indonesia (2023), pp. 125–129, https://doi.org/10.1109/IAICT59002.2023.10205656

Vasilevskyi, O., R. Trishch, V. Sarana, M. Yakovlev, and E. Popovici, Experimental analysis of blood pressure estimation using electrocardiography and photoplethysmography signals from fingertip measurements, *Acta IMEKO*, 14(2), 2025a, 1–9. https://doi.org/10.21014/actaimeko.v14i2.2096

Vasilevskyi, O., O. Voznyak, V. Didych, V. Sevastianov, O. Ruchka, and V. Rykun, Methods for constructing high-precision potentiometric measuring instruments of ion activity, *2022 IEEE 41st International Conference on Electronics and Nanotechnology (ELNANO)* (2022), pp. 247–252, https://doi.org/10.1109/ELNANO54667.2022.9927128

Vasilevskyi, O., A. Woods, and M. Cullinan, *In Situ* monitoring and quality evaluation of the powder bed fusion process of IN718 alloy powder, *IEEE Transactions on Instrumentation and Measurement*, 74, 2025b, 1–15. https://doi.org/10.1109/tim.2025.3572966

Vasilevskyi, O., A. Woods, M. Jones, and M. Cullinan, *In situ* monitoring and quality assessment of laser powder bed fusion process for 316L stainless steel, *The International Journal of Advanced Manufacturing Technology* 2025c. https://doi.org/10.1007/s00170-025-16448-0

Vasilevskyi, O., A. Woods, M. Jones, and M. Cullinan, Quantitative methodology for assessing the quality of direct laser processing of 316L steel powder using type I and type II control errors. *Electronics*, 14(7), 2025d, 1476. https://doi.org/10.3390/electronics14071476

Vasilevskyi, O. M., Calibration method to assess the accuracy of measurement devices using the theory of uncertainty, *International Journal of Metrology and Quality Engineering*, 5, 04, 403, 2014a, https://doi.org/10.1051/ijmqe/2014017

Vasilevskyi, O. M., Methods of determining the recalibration interval measurement tools based on the concept of uncertainty, *Technical Electrodynamics*, 6, 2014b, 81–88.

Vasilevskyi, O. M., V. Y. Kucheruk, Y. T. Volodarskyi, *Uncertainty of measurement, control and test results*, Kherson: OLDI-PLUS (2020): 352, ISBN: 978-966-289-374-8.

Vasilevskyi, O. M. and P.I. Kulakov, *Elements of the theory of accuracy improvement in measurement and synchronization of angular velocities of interconnected electric motor rotors*: [*Monograph*], Vinnytsia: VNTU (2011): 170.

Vasilevskyi, O. M. and A. V. Podzharenko, Evaluation of the uncertainty of rotor moment of inertia measurement by the amplitude of turning vibrations, *Visnyk of Vinnytsia Polytechnical Institute*, 4, 2010, 5–9.

Vollmer, M., Infrared thermal imaging. In *Computer vision: A reference guide*, Cham: Springer International Publishing (2021), pp. 666–670. https://doi.org/10.1007/978-3-030-63416-2_844

Zhang, P., J. Tan, Y. Tian, H. Yan et al. Research progress on selective laser melting (SLM) of bulk metallic glasses (BMGs): A review, *International Journal of Advanced Manufacturing Technology*, 118, 2022, 2017–2057. https://doi.org/10.1007/s00170-021-07990-8

Zhu, P., X. He, H. Guan, Z. Zhang, T. Zhang, and X. Qu Investigation on the attainment of high-density 316L stainless steel with selective laser sintering, *Materials* (*Basel*), 7(1), 2023, 110. https://doi.org/10.3390/ma17010110

5 Uncertainty in Dynamic Measurements

5.1 DYNAMIC PROPERTIES OF MEASURING INSTRUMENTS

Dynamic measurements, that is, measurements performed with measuring instruments (MIs) operating in a dynamic mode, are becoming increasingly widespread in engineering and scientific research. These measurements are primarily associated with the study of the behavior of physical processes in the investigated objects. Therefore, the role of dynamic measurements is particularly significant, first, in scientific fields concerned with the investigation of the structure of matter, the analysis and synthesis of new substances and materials, and the study of objects under experimental conditions; and second, in areas of engineering and production that involve the development of new technological processes and the testing of new MIs.[1,2,3,4,5]

In addition, the transfer of unit values to MIs differs in that it is necessary to transfer not only the value of the measurand but also the unit of time. Methods of time unit realization are well known and sufficiently accurate. However, it is also necessary to consider the component of dynamic measurement uncertainty that arises when fixing the time.

When performing measurements of physical quantities, the MI always operates in a transient mode, during which the output signal changes significantly with time. This phenomenon is explained by the inertial properties of MIs, since in the general case they consist of combinations of masses and springs, capacitances and inductances, or other inertial elements, which give rise to dynamic uncertainty. In addition, in digital MIs, a signal delay is observed, caused by the finite time required for conversion into a digital code.

As a result, the measurement equation that describes the static behavior of an instrument becomes invalid in the dynamic regime. In such cases, the description is replaced by differential equations that represent the dynamic relationship between the output quantity $y(t)$ and the input quantity $x(t)$ of the MI. In these differential equations, the input signals—the cause that excites the MI—are written on the right-hand side, while the output signal—the response of the MI—is written on the left-hand side:

$$a_n y^n(t) + a_{n-1} y^{n-1}(t) + \dots + a_0 y(t) =$$

$$= b_m x^m(t) + b_{m-1} x^{m-1}(t) + \dots + b_0 x(t), \quad (5.1)$$

where $x(t)$ and $y(t)$ are the input and output quantities, respectively, m and n denote the orders of derivatives, and a and b are coefficients characterizing the properties of the MI.

DOI: 10.1201/9781003742708-5

Using the differentiation operator with respect to time ($p = \frac{d}{dt}$), the general form of Equation (5.1) can be written as:

$$y(p) = S_0 \frac{b_m p^m + b_{m-1} p^{m-1} + \ldots + 1}{a_n p^n + a_{n-1} p^{n-1} + \ldots + 1} x(p) = S(p)x(p), \tag{5.2}$$

where S_0 is the static sensitivity, that is, the sensitivity to a constant input quantity (at $p = 0$), $S(p)$ is the transfer function of the MI or the operator sensitivity.

Thus, the transfer function of an MI is defined as the ratio of the transformed output quantity $y(p)$ to the transformed input quantity $x(p)$:

$$S(p) = \frac{y(p)}{x(p)} = \frac{b_m p^m + b_{m-1} p^{m-1} + \ldots + b_1 p + b_0}{a_n p^n + a_{n-1} p^{n-1} + \ldots + a_1 p + a_0}. \tag{5.3}$$

The most characteristic features of MIs are their dynamic characteristics, which are described by first-order (Table 5.1) and second-order (Table 5.2) differential equations, and in some cases by third- or higher-order equations.

TABLE 5.1
Differential Equations of First-Order Measuring Transducers

Designation	Differential Equation	Symbol
Angle measurement transducer	$J\frac{d\beta(t)}{dt} + b_0 l_0 \beta(t) = H_0 \psi_0(t)$	J is the moment of inertia; β is the angular displacement; b_0 is the specific damping force; l_0 is the distance from the axis of rotation to the line of action of the damping force; H_0 is the angular momentum vector; $\psi_0(t)$ is the instantaneous value of the angular displacement of the object.
Gas humidity sensor	$\frac{d\varphi_r(t)}{dt} + \frac{1}{\lambda_0}\varphi_r(t) = \frac{1}{\lambda_0}\varphi_\bullet(t)$	$\phi_r(t)$ is the hygrometer reading; λ_0 is the coefficient characterizing the measurement conditions and the structural features of the hygrometer; $\phi_\bullet(t)$ is the instantaneous value of the gas relative humidity (measurand).
Flow rate transducer	$J\frac{d\omega(t)}{dt} + r_0\omega(t) = c_0 \vartheta^2(t)$	J is rotor moment of inertia; $\omega(t)$ is the rotor angular velocity; r_0 is the friction coefficient; c_0 is the anemometer constant; $\theta(t)$ is the flow velocity.
Thermal transducer	$\frac{dU(t)}{dt} + \frac{\alpha_k S}{mc}U(t) = \frac{\alpha_k S}{mc}\theta(t)$	$U(t)$ is the sensor temperature; S is the sensor surface area; m is the sensor mass; c is the specific heat capacity of the sensor material; α_k is the convective heat transfer coefficient; $\theta(t)$ is the measured temperature (measurand).

(*Continued*)

Table 5.1 (Continued)

Designation	Differential Equation	Symbol
Flow meter transducer	$J\frac{d\omega(t)}{dt} + DQ\omega(t) = AQ^2$	J is the moment of inertia of the impeller; ω is the impeller angular velocity; Q is the volumetric flow rate; A, D are constant coefficients.
Pressure measurement transducer	$\tau_0 \frac{dP_2(t)}{dt} + P_2(t) = P_1(t)$; $\tau_0 = \frac{128\mu_0 V_0 l_0}{\pi d^4 P_2}$.	$P_2(t)$ is the instantaneous value of the gas pressure; $P_1(t)$ is the pressure at the inlet of the manometer tube; μ_0 is the dynamic viscosity coefficient of the gas; V_0 is the gas volume; l_0 and d are the length and the diameter of the manometer tube, respectively.

TABLE 5.2
Differential Equations of Second-Order Measuring Transducers

Designation	Differential Equation	Symbol
Body acceleration transducer	$m\frac{d^2x_0(t)}{dt^2} + k_1\frac{dx_0(t)}{dt} + c_1x_0(t) = ma(t)$	$a(t)$ is the measured acceleration; $x_0(t)$ is the displacement of the instrument's inertial mass m; k_1, c_1 are stiffness and damping coefficients of the elastic element, respectively.
Angular velocity transducer	$J\frac{d^2\varphi(t)}{dt^2} + b_0l_0\frac{d\varphi(t)}{dt} + c_0l_1^2\varphi(t) = H_0\omega(t)$	$\phi(t)$ is the angular displacement; c_0 is the stiffness of the restoring spring; l_1 is the distance from the output axis to the line of action of the spring force; $\omega(t)$ is the angular velocity (measurand).
Gas humidity transducer	$a_0T_0^2\frac{d^2\varphi_r(t)}{dt^2} + (a_0T_0 + T_0)\frac{d\varphi_r(t)}{dt} + {} + \varphi_r(t) = k\varphi_u(t)$	k is the scale factor; a_0, T_0 are parameters determined experimentally.
Thermal transducer	$\frac{d^2U(t)}{dt^2} + (\beta_1 + \beta_2 + \beta_3\alpha_k)\frac{dU(t)}{dt} + {} + \beta_1\beta_3\alpha_kU(t) = \beta_1\beta_3\alpha_k\theta(t)$	$U(t)$ is the thermometer reading; $\beta_1 = k_0S_e/c_e$, $\beta_2 = k_0S_e/c_r$, $\beta_3 = S_r/c_r$; c_e, c_r are the total heat capacities of the sensor and the casing; k_0 is the heat transfer coefficient; S_e, S_r are surface areas of the sensor and the casing.
Pressure measurement transducer	$m\frac{d^2W(t)}{dt^2} + k_1\frac{dW(t)}{dt} + c_1W(t) = P(t)$	m is the membrane mass; k_1 is the damping coefficient; c_1 is the membrane stiffness; $W(t)$ is instantaneous membrane deflection; $P(t)$ is the measured pressure (measurand).

Information about the dynamic characteristics should be provided in the normative and technical documentation of the MI.

For expressing the dynamic uncertainty of measurement results, the frequency response of the instrument is convenient for practical applications. The frequency response $S(j\omega)$ is obtained by substituting the operator p in the transfer function $S(p)$ (Equation 5.3) with $j\omega$.

For a non-inertial MI (ideal transducer), the frequency response can be represented as:

$$S(j\omega) = K, \tag{5.4}$$

where K is the transfer coefficient of the MI.[6]

For MIs with a first-order (aperiodic) element (e.g., resistance temperature transducers), the frequency response is given by:

$$S(j\omega) = \frac{K}{1 + j\omega\tau}, \tag{5.5}$$

where τ is the time constant, determined by the parameters of the MI.

For MIs with an integrating element (e.g., integrating amplifiers), the frequency response is given by:

$$S(j\omega) = \frac{K}{j\omega}. \tag{5.6}$$

For MIs with a derivative (lead) element (e.g., differentiating amplifiers), the frequency response is given by:

$$S(j\omega) = K(1 + j\omega\tau). \tag{5.7}$$

For MIs with a time-delay element (e.g., analog-to-digital converters), the frequency response is given by:

$$S(j\omega) = \exp(-j\omega\tau). \tag{5.8}$$

For MIs with an oscillatory element (e.g., electromechanical measuring transducers), the frequency response is given by:

$$S(j\omega) = \frac{K}{1 + j\omega\tau_1 - \omega^2\tau_2^2} = \frac{K}{1 + 2j\omega\beta\tau - \omega^2\tau^2}. \tag{5.9}$$

5.2 REPRESENTATION OF UNCERTAINTY IN DYNAMIC MEASUREMENTS

Dynamic measurement uncertainty refers to the component of measurement uncertainty that arises from the response of the MI to the frequency (rate) of change of the input signal $X(t)$. This component depends on the dynamic properties of the MI and the frequency spectrum of the input signal.[7]

If the transfer equation of the MI operating in the dynamic regime can be represented in the form:

$$Y(t) = KX(t), \tag{5.10}$$

where $X(t)$ is the value of the physical quantity being measured (input signal), K is the transfer coefficient of the MI, and $Y(t)$ is the measurement result (output signal).

Then, the mathematical expectation of the input signal will be equal to M[X], and the mathematical expectation of the output signal will be equal to

$$M[Y] = KM[X], \tag{5.11}$$

where $M[Y]$ and $M[X]$ are, respectively, the mathematical expectations of the output and input signals of the MI.

The spectral density of the input signal $X(t)$ is given as follows:

$$H_X(\omega) = \lim\,(2T)^{-1}|X(j\omega)|^2 \text{ for } T \to \infty, \tag{5.12}$$

where $X(j\omega)$ is the Fourier transform obtained by substituting $s = j\omega$ into the Laplace transform representation of the signal $X(s)$, T is the observation time, $\omega = 2\pi f$.

The expression for the spectral density of the output signal can be represented as follows:

$$H_Y(\omega) = \lim\,(2T)^{-1}|Y(j\omega)|^2 \text{ for } T \to \infty. \tag{5.13}$$

The ratio of the output to the input representations yields the expression for describing the transfer function of the MI:

$$K(s) = \frac{Y(s)}{X(s)} = \frac{\sum_{k=0}^{m} B_k s^k}{\sum_{q=0}^{n} A_q s^q}, \tag{5.14}$$

where $Y(s)$ and $X(s)$ are the Laplace transform representations of the output signal $Y(t)$ and the input signal $X(t)$, respectively; k and q are the orders of derivatives of Y and X, respectively; and A_q and B_k are the coefficients of the differential equation:

$$H_{\Delta X}(\omega) = \left|\frac{K(j\omega)}{K} - 1\right|^2 H_X(\omega), \tag{5.15}$$

where K is the frequency response of the MI at $\omega = 0$.

The power spectral density of the output signal is described by the following expression

$$H_Y(\omega) = |K(j\omega)|^2 H_X(\omega), \tag{5.16}$$

where $K(j\omega)$ is the frequency response of the MI.

The standard deviation (dynamic measurement uncertainty) of the output signal Y in the frequency domain during dynamic measurements can be estimated as the square root of the integral of the product of the squared modulus of the frequency response of the MI and the squared spectral function of the input signal over a wide frequency range:

$$\sigma_Y(\omega) = \pi^{-1/2}\left(\int_0^{\pi Fs} |K(j\omega)|^2 H_X(\omega) d\omega\right)^{1/2} = \\ = \pi^{-1/2}\left((2\mathrm{T})^{-1}\int_0^{\pi Fs} |K(j\omega)|^2 |X(j\omega)|^2 d\omega\right)^{1/2}, \tag{5.17}$$

where $|K(j\omega)|$ is the modulus of the frequency response of the MI, which can be used as a mathematical model of the instrument in dynamic measurements for the evaluation of dynamic measurement uncertainty, and Fs is the sampling frequency.[8]

Similarly, the standard deviation of the dynamic error of the MI in the frequency domain—which gives rise to the dynamic component of measurement uncertainty, referred to the input of the instrument—is described by the following expression:

$$\sigma_{\Delta X}(\omega) = \pi^{-1/2}\left((2\mathrm{T})^{-1}\int_0^{\pi Fs} \left|\frac{K(j\omega)}{K} - 1\right|^2 |X(j\omega)|^2 d\omega\right)^{1/2}. \tag{5.18}$$

The modulus of the frequency response of the MI can be obtained using the following expression:

$$|K(j\omega)| = \left(a^2(\omega) + b^2(\omega)\right)^{1/2}, \tag{5.19}$$

where $a(\omega)$, $b(\omega)$ are the real and imaginary parts of the frequency response of the MI, respectively.

The spectral function of the input quantity $X(j\omega)$ is related to its time-domain representation $X(t)$ via the Laplace transform:

$$\mathrm{X(j\omega)} = \int_0^{\infty} X(t)e^{-j\omega t} dt, \tag{5.20}$$

where ω is the angular frequency.

For a finite observation interval, the infinity sign can be replaced by the total observation time T. To represent the dynamic measurement uncertainty in the time domain, $u_D(t)$, the inverse Fourier transform should be applied as follows:

$$u_D(t) = \pi^{-1/2}\int_0^{\pi Fs} \sigma_{\Delta X}(\omega)e^{j\omega t} d\omega = \\ = \pi^{-1/2}\left[\int_0^{\pi Fs} \sigma_{\Delta X}(\omega)\cos(\omega t) d\omega + j\int_0^{\pi Fs} \sigma_{\Delta X}(\omega)\sin(\omega t) d\omega\right]. \tag{5.21}$$

Since expression (5.21) consists of real and imaginary parts, and for the evaluation of uncertainty we are primarily interested in the amplitude (maximum) value of the dynamic measurement uncertainty, expression (5.21) can be represented in terms of its modulus:

$$|u_D(t)| = \left[\left(\int_0^{\pi Fs} \pi^{-1/2}\sigma_{\Delta X}(\omega)\cos(\omega t)d\omega\right)^2 + \left(\int_0^{\pi Fs} \pi^{-1/2}\sigma_{\Delta X}(\omega)\sin(\omega t)d\omega\right)^2\right]^{1/2}. \tag{5.22}$$

Thus, the amplitude of the dynamic measurement uncertainty arising from the inertial properties of the MI used during dynamic measurements can be evaluated in the time domain based on the model equation of the input quantity's spectral function and the frequency response of the MI, as given by expression (5.22).

If the measured quantity $x(t)$ is discrete, the integral sign in Equation (5.20) can be replaced by a summation sign, with the following substitutions: t is replaced by nT_a, where n ranges from 0 to $N-1$, and T_a denotes the sampling period. Then $x(t)s$ is represented as $x(nT_a)$, and the corresponding differential element $e^{-j\omega_0 t}$ is replaced by $e^{-j\omega_0 nT_a}$.

After these substitutions, Equation (5.20) can be written in discrete form

$$X_d(j\omega) = \sum_{n=0}^{N-1} x(nT_a)e^{-j\omega_0 nT_a} = \sum_{n=0}^{N-1} x(nT_a)\cos\omega_0 nT_a - j\sum_{n=0}^{N-1} x(nT_a)\sin\omega_0 nT_a, \tag{5.23}$$

where $\omega_0 = \frac{2\pi}{NT_a}k$ and $k = 0, 1, \ldots, N-1$.

In order for the discrete spectral function to correspond in magnitude to the continuous spectral function, it should be multiplied by the sampling interval.

$$X(j\omega) = T_a X_d(j\omega). \tag{5.24}$$

For dynamic measurements of time-discrete signals, the equation for evaluating the dynamic measurement uncertainty of the output quantity YYY in the frequency domain (5.17), considering Equations (5.23) and (5.24), can be expressed as follows:

$$u_D = \sqrt{\frac{T_a}{N}\sum_{k=0}^{N-1}\sum_{n=0}^{N-1} x^2(nT_a)e^{-j\frac{4\pi nk}{N}}A^2\left(k\frac{2\pi}{NT_a}\right)}, \tag{5.25}$$

where $A\left(k\frac{2\pi}{NT_a}\right) = A(\omega) = |S(j\omega)|$ is the amplitude–frequency response (AFR) of the MI, $\Delta\omega = \frac{2\pi}{NT_a}$ is the spacing between discrete frequency values, T_a is the sampling period, N is the number of samples, and NT_a is the total observation time.

5.3 EVALUATION OF DYNAMIC MEASUREMENT UNCERTAINTY IN VIBRATION ACCELERATION MEASUREMENTS

When measuring vibration parameters of operating machinery, the mechanical vibrations acting on the accelerometer base cause a dynamic force on the piezoelectric element, equal to the product of the acceleration of the seismic mass and its corresponding mass. The piezoelectric elements generate an electric charge proportional to the dynamic force acting on them. The seismic masses of the accelerometer are practically constant, which means that the electric charge generated on the electrodes of the piezoelectric elements is proportional to the acceleration of these masses.

Since the amplitude and phase of the acceleration of the seismic masses over a wide frequency range are identical to the amplitude and phase of the acceleration of the accelerometer base, the total electric charge delivered by the piezoelectric elements is proportional to the acceleration of the base and, consequently, to the acceleration of the mechanical vibrations of the surface of the object to which the accelerometer is attached.

The differential equation describing the dynamic relationship between the output and input quantities of the vibration acceleration measurement channel is given by:

$$m\frac{d^2X_s(t)}{dt^2} + c\frac{dX_s(t)}{dt} + kX_s(t) = F(t), \tag{5.26}$$

where $F(t) = F_0 \sin(\omega_0 t)$ is the harmonic excitation force of the object surface vibrations (input quantity), F_0 is the amplitude of the force, ω_0 is the angular frequency of the excitation force, $X_s(t)$ is the mechanical displacement of the inertial mass, m is the mass of the accelerometer, c is the damping coefficient, and k is the equivalent stiffness of the piezoelectric elements.

The differential equation (5.26) can be expressed in the commonly used form as follows:

$$\frac{d^2X_s(t)}{dt^2} + 2h\frac{dX_s(t)}{dt} + h_{cr}^2 X_s(t) = \frac{F_0}{m}\sin(\omega_0 t), \tag{5.27}$$

where $h = c/(2m)$ is the damping coefficient, and $h_{cr} = \sqrt{k/m}$ is the critical value of the damping coefficient.

The transfer function of this MI is given by:

$$S(p) = \frac{K}{m\left(p^2 + 2hp + h_{cr}^2\right)}, \tag{5.28}$$

where K is the proportionality coefficient of the vibration acceleration measurement channel.

By transforming to the frequency domain and separating the real and imaginary parts, the modulus of the frequency response can be expressed as follows:

$$|S(j\omega)| = \frac{K}{m|p^2 + 2hp + h_{cr}^2|}. \tag{5.29}$$

The input quantity $F_0 \sin(\omega_0 t)$, according to Equation (5.20), has the following representation:

$$X(j\omega) = F_0 \frac{\omega_0}{\omega_0^2 + (j\omega)^2}, \tag{5.30}$$

where ω_0 is the angular frequency of the input quantity.

The amplitude of the harmonic excitation force F_0 is approximately 0.1 mm, corresponding to 10^{-4} m. The mass of the accelerometer mmm is 40 g, corresponding to $4{\cdot}10^{-2}$ kg. The damping coefficient ccc for piezoelectric accelerometers is 0.5, and the equivalent stiffness of the piezoelectric elements is taken as $k = 2$. The proportionality coefficient K of the vibration acceleration measurement channel is 100.

By substituting the calculated values of the modulus of the frequency response (5.29) and the representation of the input (measured) quantity into the equation for evaluating the dynamic measurement uncertainty of the output quantity in the frequency domain, we obtain the representation of the dynamic uncertainty in the form:

$$u_D(j\omega) = \left[\frac{1}{2\pi}\int_{-\infty}^{+\infty} \frac{K^2}{m^2\left|(j\omega)^2 + j2h\omega + h_{cr}^2\right|^2}\left|\frac{F_0\omega_0}{\omega_0^2 + (j\omega)^2}\right|^2 d\omega\right]^{0.5}. \tag{5.31}$$

To determine the numerical value of the representation of the dynamic measurement uncertainty (5.31), the integral under the square root can be expressed as follows:

$$I_n = \frac{1}{2\pi}\int_{-\infty}^{\infty} \frac{G(j\omega)}{K(j\omega)K(-j\omega)} d\omega, \tag{5.32}$$

where $K(j\omega) = a_0(j\omega)^n + a_1(j\omega)^{n-1} + \ldots + a_n$, $G(j\omega) = b_0(j\omega)^{2n-2} + b_1(j\omega)^{2n-4} + \ldots + b_{n-1}$.

The polynomial contains only even powers of $G(j\omega)$.

For $n = 1$

$$I_1 = \frac{1}{2\pi}\int_{-\infty}^{\infty} \frac{b_0}{\left|a_0 \cdot j\omega + a_1\right|^2} d\omega = \frac{b_0}{2a_0a_1}. \tag{5.33}$$

For $n = 2$

$$I_2 = \frac{1}{2\pi}\int_{-\infty}^{\infty} \frac{b_0 \cdot (j\omega)^2 + b_1}{\left|a_0 \cdot (j\omega)^2 + a_1 \cdot j\omega + a_2\right|^2} d\omega = \frac{-b_0 + \frac{a_0b_1}{a_2}}{2a_0a_1}. \tag{5.34}$$

For $n = 3$

$$I_3 = \frac{1}{2\pi}\int_{-\infty}^{\infty} \frac{b_0 \cdot (j\omega)^4 + b_1 \cdot (j\omega)^2 + b_2}{\left|a_0 \cdot (j\omega)^3 + a_1 \cdot (j\omega)^2 + a_2 \cdot j\omega + a_3\right|^2} d\omega = \frac{-a_2b_0 + a_0b_1 - \frac{a_0a_1b_2}{a_3}}{2a_0(a_0a_3 - a_1a_2)}. \tag{5.35}$$

For $n = 4$

$$I_4 = \frac{1}{2\pi}\int_{-\infty}^{\infty} \frac{b_0\cdot(j\omega)^6 + b_1\cdot(j\omega)^4 + b_2\cdot(j\omega)^2 + b_3}{\left|a_0\cdot(j\omega)^4 + a_1\cdot(j\omega)^3 + a_2\cdot(j\omega)^2 + a_3\cdot j\omega + a_4\right|^2}d\omega =$$

$$= \frac{b_0(-a_1a_4 + a_2a_3) - a_0a_3b_1 + a_0a_1b_2 + \frac{a_0b_3}{a_4}(a_0a_3 - a_1a_2)}{2a_0\left(a_0a_3^2 + a_1^2a_4 - a_1a_2a_3\right)}. \quad (5.36)$$

The integral under the square root in the representation of the dynamic measurement uncertainty (5.31) can be evaluated using the integral in Equation (5.36). In Equation (5.36), after some transformations of Equation (5.31), the numerator coefficients are: $b_0 = 0$, $b_1 = 0$, $b_2 = 0$, $b_3 = \frac{K^2F_0^2\omega_0^2}{m^2}$, and the denominator coefficients are: $a_0 = 1$, $a_1 = 2h$, $a_2 = h_{cr}^2 + \omega_0^2$, $a_3 = 2h\omega_0^2$, and $a_4 = h_{cr}^2\omega_0^2$.

By substituting the above coefficients and their numerical values into the right-hand side of Equation (5.36) and taking the square root, we obtain the numerical solution for the dynamic measurement uncertainty referred to the output of the MI (5.31), which is equal to 0.02 m/s^2.

The uncertainty budget of the elements of the vibration acceleration measurement channel in relative units is presented in Table 5.3. Based on the experimental study of the uncertainty budget of the vibration acceleration measurement channel (Table 5.3), the relative value of the combined standard uncertainty, excluding the dynamic component, is 0.36%.

TABLE 5.3
Uncertainty Budget of the Components of the Vibration Acceleration Measurement Channel

Elements	Value of Relative Uncertainty, %	The Expanded Uncertainty (Coverage Factor 1.96 at Confidence Level 0.95), %	Distribution
Accelerometer	0.3	0.59	Uniform (rectangular)
Preliminary charge amplifier	0.02	0.04	Uniform (rectangular)
Bandpass filter	0.2	0.39	Uniform (rectangular)
Scale converter	0.01	0.02	Uniform (rectangular)
ADC	0.005	0.01	Uniform (rectangular)
Background noise	$2.14\cdot10^{-6}$	$4.19\cdot10^{-6}$	Normal
The combined standard uncertainty ($\tilde{u}_c$)	0.36	0.71	Normal
Mean value			2.93 m/s^2

The accelerometer is mounted on the motor's electric spindle and performs point measurements of vibrations in the frequency range from 6 to 10 kHz, generated by the operation of rolling bearings. Vibrations induced by other excitation forces should be considered as disturbances.

For the evaluation of the dynamic measurement uncertainty referred to the input of the MI, the transfer function (5.29) must be expressed in the frequency domain, and the real and imaginary parts separated. As a result of these transformations, the modulus of the frequency response of the vibration acceleration MI is obtained, which is described by the following expression:

$$|S(j\omega)| = \left|\frac{K}{(j\omega)^2 + 2h(j\omega) + h_{cr}^2}\right| = \left[\frac{K_{\text{ВП}}}{\omega^4 - 2\omega^2 h_{cr}^2 + 4\omega^2 h^2 + h_{cr}^4}\right]^{1/2}, \quad (5.37)$$

as well as the frequency response of the MI at $\omega = 0$, $S(0) = \frac{mK}{k}$.

The modulus of the representation of the input vibration acceleration quantity is given by:

$$|X(j\omega)| = \omega_0 F_0 \left(\omega_0^2 - \omega^2\right)^{-1} m^{-1}, \quad (5.38)$$

where ω_0 is the angular frequency of the input vibration acceleration quantity, varying in the range from 6 to 10 kHz; that is, its minimum value is 18849.5 rad/s, and its maximum value is 31415.9 rad/s.[9]

By substituting the calculated values of the modulus of the frequency response and the representation of the input quantity into the equation for evaluating the dynamic measurement uncertainty referred to the input of the MI in the frequency domain (5.18), we obtain:

$$\sigma_{\Delta X}(\omega) = \pi^{-1/2}\left(\frac{1}{2\text{T}}\int_0^{\pi F_s}\frac{\omega F_0\left[\begin{array}{r}\left(k\left(h_{cr}{}^2 - \omega^2\right) - m\left(h_{cr}{}^2 - \omega^2\right)^2 - 4m\omega^2 h^2\right)^2 \\ + 4k^2\omega^2 h^2\end{array}\right]}{2Tm\left(\omega_0^2 - \omega^2\right)\left(m\left(h_{cr}{}^2 - \omega^2\right)^2 + 4m\omega^2 h^2\right)^2}d\omega\right)^{1/2}. \quad (5.39)$$

To represent the time-domain behavior of the dynamic measurement uncertainty of the vibration acceleration MI, which is caused by the inertial properties of the instrument under dynamic operating conditions, the inverse Fourier transform described by expression (5.21) must be applied.

Since expression (5.21) consists of real and imaginary parts, and we are primarily interested in the amplitude of the dynamic measurement uncertainty in the time domain (5.22), the expression for evaluating the dynamic measurement uncertainty, considering the previously obtained representation of the dynamic uncertainty in the frequency domain, can be written as follows:

$$|u_D(t)| = \left[\left(\int_0^{\pi F_s} \pi^{-1/2}\sigma_{\Delta X}(\omega)\cos(\omega t)d\omega\right)^2 + \right.$$
$$\left. + \left(\int_0^{\pi F_s} \pi^{-1/2}\sigma_{\Delta X}(\omega)\sin(\omega t)d\omega\right)^2\right]^{0.5}. \quad (5.40)$$

The solution of the obtained equation for the modulus of the dynamic measurement uncertainty in the time domain was performed using the Maple 12 software package. By substituting the values of the influencing coefficients mentioned above into this expression, the amplitude of the dynamic measurement uncertainty of vibration acceleration was determined to be 0.068 m/s^2 (Figure 5.1) for an input vibration acceleration frequency of 10 kHz and an observation time of $T = 300$ s. If the observation time is increased to 600 s at the same input signal frequency, the dynamic uncertainty decreases to 0.049 m/s^2 (Figure 5.2).

At the lower input signal frequency of 6 kHz, with an observation time of 300 s, the dynamic measurement uncertainty is 0.182 m/s^2 (Figure 5.3). If the observation time is increased to 600 s at the input signal frequency of 6 kHz, the dynamic

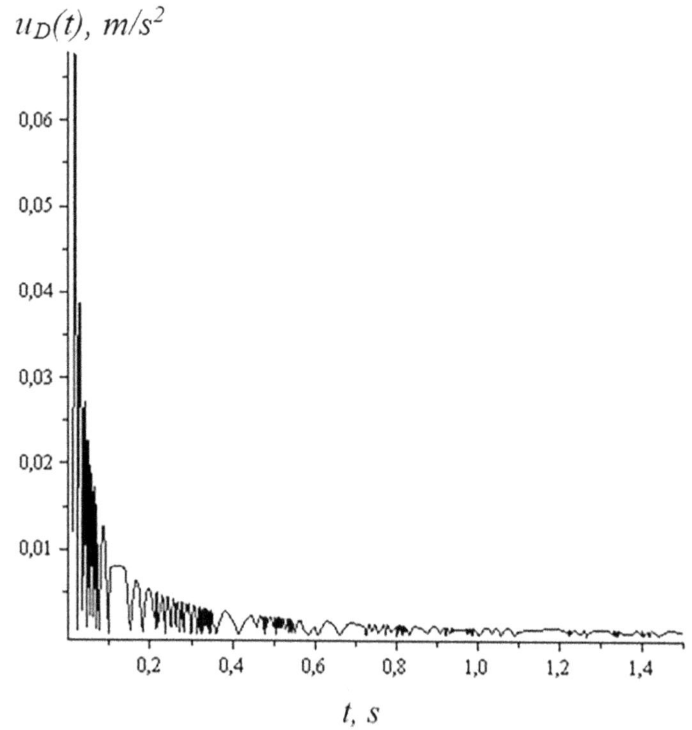

FIGURE 5.1 Time-domain characteristic of the dynamic measurement uncertainty of the vibration acceleration MI at an input frequency of 10 kHz and an observation time of 300 s.

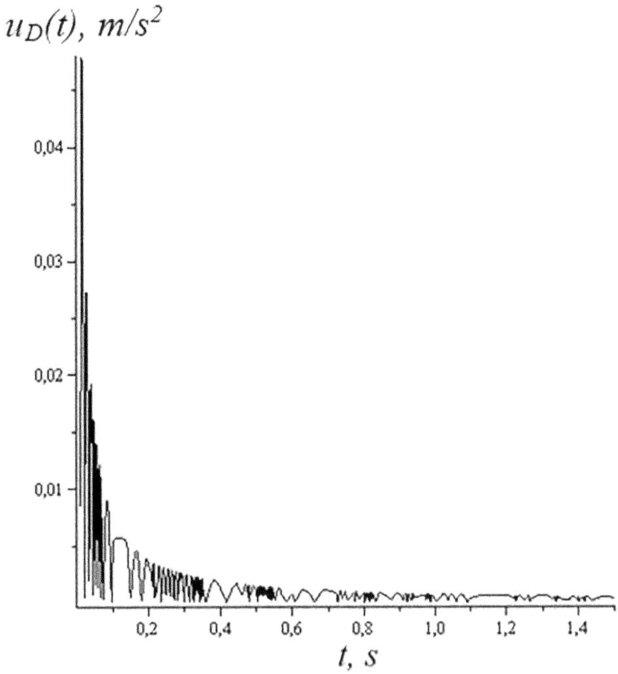

FIGURE 5.2 Time-domain characteristic of the dynamic measurement uncertainty of the vibration acceleration measuring instrument at an input frequency of 10 kHz and an observation time of 600 s.

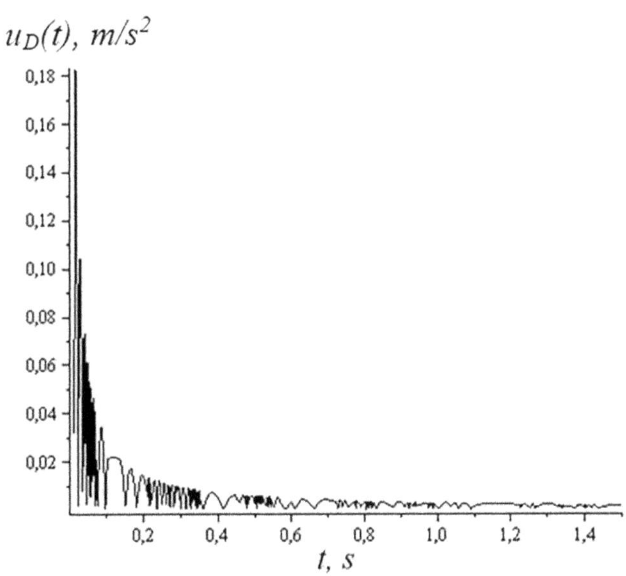

FIGURE 5.3 Time-domain characteristic of the dynamic measurement uncertainty of the vibration acceleration measuring instrument at an input frequency of 6 kHz and an observation time of 300 s.

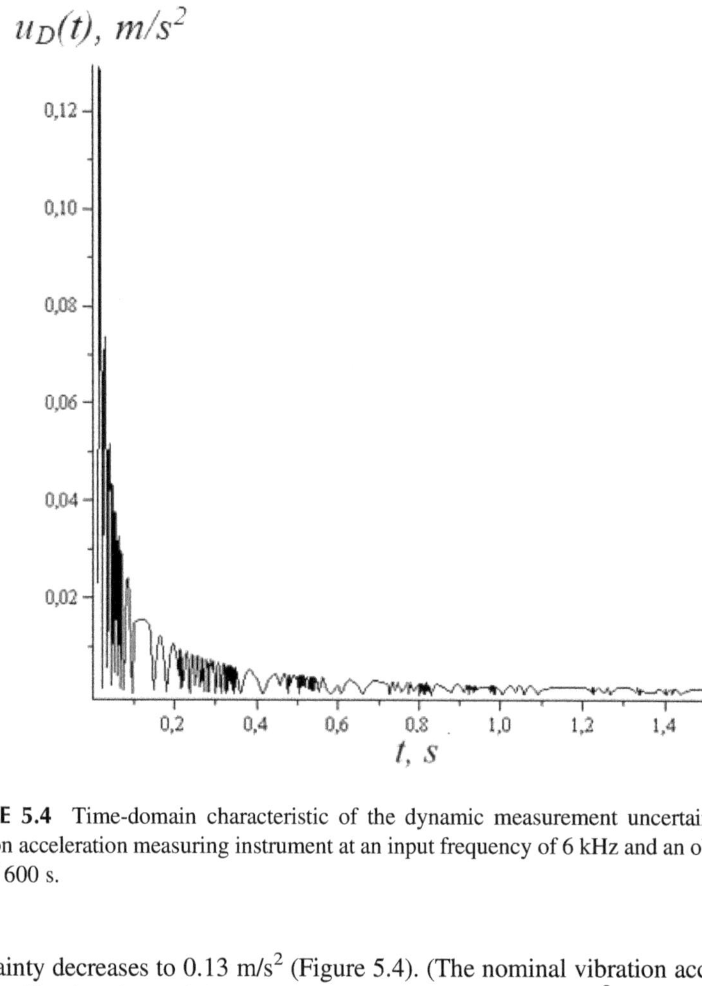

FIGURE 5.4 Time-domain characteristic of the dynamic measurement uncertainty of the vibration acceleration measuring instrument at an input frequency of 6 kHz and an observation time of 600 s.

uncertainty decreases to 0.13 m/s^2 (Figure 5.4). (The nominal vibration acceleration of the rolling bearings of the motor's electric spindle is 2.93 m/s^2).

The time-domain characteristics of the dynamic measurement uncertainty of vibration acceleration, obtained using the Maple 12 software package, are presented in Figures 5.1–5.4 for the minimum and maximum values of the input signal frequency (6 and 10 kHz) with observation times of 300 and 600 s, respectively.

Thus, based on the proposed method for evaluating dynamic measurement uncertainty, the maximum value of the dynamic uncertainty was calculated using the model equation of a vibration transducer. The evaluation was performed using mathematical models of the spectral function of the input vibration acceleration signal and the frequency response of the vibration transducer. This approach allows, in addition to the values of static standard uncertainties presented in Table 5.4, the consideration of dynamic uncertainty arising under dynamic operating conditions of the MI when calculating the combined standard uncertainty of the measurement.

To calculate the maximum value of the relative dynamic measurement uncertainty of vibration acceleration, the obtained maximum value of the dynamic uncertainty,

TABLE 5.4
Dynamic Measurement Uncertainty Budget of Vibration Acceleration Referred to the Input of the Measuring Instrument

Quantity	Mean Value, m/s^2	Frequency of the Study, kHz	Observation Time, s	Value of Dynamic Uncertainty, m/s^2	The Expanded Dynamic Uncertainty (Coverage Factor 1.96 at Confidence Level 95%), m/s^2	Value of Relative Dynamic Uncertainty, %
Vibration acceleration	2.93	6	300	0.182	0.36	6.21
		6	600	0.13	0.25	4.44
		10	300	0.068	0.13	2.32
		10	600	0.049	0.1	1.67

0.182 m/s^2 (Figure 5.3), is divided by the estimated vibration acceleration $\overline{X} = 2.93\ \mathrm{m/s^2}$, resulting in:

$$\tilde{u}_D = \frac{u_D(t)}{\overline{X}} 100\% = \frac{0.182}{2.93} 100\% = 6.21\%. \tag{5.41}$$

When calculating the relative dynamic measurement uncertainty of vibration acceleration at a frequency of 6 kHz and an observation time of 600 s, a value of 4.44% is obtained (see Figure 5.4). At a frequency of 10 kHz and an observation time of 300 s (see Figure 5.1), the relative dynamic measurement uncertainty is 2.32%. At a frequency of 10 kHz and an observation time of 600 s (see Figure 5.2), the relative dynamic measurement uncertainty is 1.67%.

The dynamic measurement uncertainty budget of vibration acceleration for observation times of 300 600 s at frequencies of 6 and 10 kHz is presented in Table 5.4. The maximum value of dynamic uncertainty is 6.21% at a frequency of 6 kHz and an observation time of 300s (Table 5.4).

The relative combined standard uncertainty of vibration acceleration measurement, considering the values of the relative static combined standard uncertainty (Table 5.3) and the relative dynamic uncertainty (Table 5.4), can be calculated using the following expression:

$$\tilde{u}_C = \sqrt{\tilde{u}_D^2 + \tilde{u}_c^2} = \sqrt{6.21^2 + 0.36^2} = 6.22\%. \tag{5.42}$$

The maximum relative combined standard uncertainty of vibration acceleration measurement is 6.22% for an observation time of 300 s at a frequency of 6 kHz. When the observation time is increased to 600 s at the same frequency, the combined uncertainty decreases to 4.45%. At a frequency of 10 kHz, the combined uncertainty is 2.35% for an observation time of 300 s, and 1.71% for an observation time of 600 s.

5.4 EVALUATION OF MEASUREMENT UNCERTAINTY OF THE DYNAMIC TORQUE OF ROTATING SYSTEMS

The differential equation describing the dynamic relationship between the output and input quantities during the operation of a MI for the dynamic torque of a rotating system (RS) has the following form:

$$\frac{d^2\varphi(t)}{dt^2} + 2\upsilon\omega_p\frac{d\varphi(t)}{dt} + \omega_p^2\varphi(t) = \frac{M_c(t)}{J_c}, \tag{5.43}$$

where J_C is the total moment of inertia of the stator of the rotating system and the moving part of the measuring transducer, $\upsilon = \frac{P}{2\sqrt{J_c C}}$ is the damping ratio of the free oscillations of the measuring transducer, $\omega_p = \sqrt{C/J_c}$ is the natural frequency of the undamped free oscillations of the measuring transducer, C is the stiffness of the force sensor, and P is the damping coefficient.[10]

The transfer function of this MI for dynamic torque is described by the following expression:

$$S(p) = \frac{K}{p^2 + 2v\omega_p p + \omega_p^2}, \tag{5.44}$$

where $K = \frac{gK1}{J_c\omega_p^2}$ is the proportionality coefficient of the dynamic torque MI, g is the acceleration due to gravity, and $K1$ is a constant of the strain-gauge transducer.

By transforming to the frequency domain and separating the real and imaginary parts, as well as performing the corresponding mathematical manipulations, we obtain the following expressions for the frequency response of the dynamic torque MI and for the square of the magnitude of its frequency response, respectively:

$$S(j\omega) = \frac{K}{-\omega^2 + j2v\omega_p\omega + \omega_p^2}; \tag{5.45}$$

$$|S(j\omega)|^2 = \frac{K^2\left(\omega^4 + 4v^2\omega^2\omega_p^2 - 2\omega^2\omega_p^2 + \omega_p^4\right)}{\begin{matrix}\omega^8 + 4\omega^6\omega_p^2(2v^2 - 1) + 2\omega^4\omega_p^4(3 - 8v^2) \\ + 8v^2\omega^2\omega_p^4\left(2v^2\omega^2 + \omega_p^2\right) - 4\omega^2\omega_p^6 + \omega_p^8\end{matrix}}. \tag{5.46}$$

The frequency response of the measuring transducer at $\omega = 0$ is equal to $S(0) = \frac{K}{\omega_p^2}$.

The input signal (the right-hand side of the differential Equation (5.43), based on a priori information obtained during the investigation of the dynamic torque characteristics of the RS, can be described by the following mathematical model:

$$M_c(t) = M_k e^{-v\omega_0 t}, \tag{5.47}$$

Where M_k is the value of the torque at the moment of stator power-off of the RS, and ω_0 is the angular frequency of the input signal.

The transfer function of the input signal (5.47) is given by:

$$X(p) = \frac{M_k}{p + \nu\omega_0}. \tag{5.48}$$

Transforming to the frequency domain, the spectral function of the input signal can be expressed as:

$$X(j\omega) = \frac{M_k}{j\omega + \nu\omega_0}. \tag{5.49}$$

The square of the magnitude of the spectral function of the input signal, after separating the real and imaginary parts and performing the corresponding mathematical transformations, can be expressed as:

$$|X(j\omega)|^2 = \frac{M_k^2\left(\omega^2 + \nu^2\omega_0^2\right)}{\omega^4 + 2\omega^2\nu^2\omega_0^2 + \nu^4\omega_0^4}. \tag{5.50}$$

By substituting the obtained expressions for the square of the magnitude of the spectral function of the input signal (5.50) and the square of the magnitude of the frequency response of the dynamic torque MI (5.46) into expression (5.18) for representing the dynamic uncertainty referred to the input, we obtain:

$$\sigma_{\Delta X}(\omega) = \left(\frac{1}{2T\pi}\int_0^{\pi F_s} \frac{M_k^2\left(\omega^2 + \nu^2\omega_0^2\right)}{\omega^4 + 2\omega^2\nu^2\omega_0^2 + \nu^4\omega_0^4} d\omega\right)^{0.5} \times$$
$$\times \left(\int_0^{\pi F_s} \frac{\left(-\omega^4 - 4\nu^2\omega^2\omega_p^2 + \omega^2\omega_p^2\right)^2 + 4\nu^2\omega^2\omega_p^6}{\left(\omega^4 - 2\omega^4\omega_p^2 + 4\nu^2\omega^2\omega_p^2 + \omega_p^4\right)^2} d\omega\right)^{0.5}. \tag{5.51}$$

To solve the above expression (5.51) and construct the characteristics of the dynamic uncertainty variation, the Maple software package was used.

Since the methodology for measuring the dynamic torque of the RS involves deenergizing the stator of the RS after it reaches nominal operating conditions and subsequently measuring the dynamic torque in a self-braking mode, during which the rotational frequency decreases from the nominal value of 50 Hz to a complete stop of the rotor, we investigate how the dynamic uncertainty characteristics change (i.e., the values they assume) at the beginning of the measurements (upper measurement limit) at the nominal frequency of 50 Hz and at the lower measurement limit as the input signal frequency decreases to 1 Hz.

To represent the dynamic uncertainty variation in the time domain, it is necessary to perform an inverse Fourier transform and determine the magnitude of this uncertainty, which is described by expression (5.22), considering the expression of the dynamic uncertainty referred to the input $\sigma_{\Delta X}(\omega)$. By substituting the nominal numerical values of the quantities involved in the frequency-domain dynamic uncertainty equation, $\sigma_{\Delta X}(\omega)$ is g=9.81 m/s^2, $K1$=489.89 N, J_c=0.02 N · m^2, M_k=8 N·m, C=3000 N·m/rad, and P=0.75 N·m · s/rad—we obtain the characteristics of dynamic uncertainty variation during the measurement of dynamic torque in the range from 8 to 0.5 N·m at the nominal rotor speed of 50 Hz (Figure 5.5) and at the minimum rotor speed of 1 Hz (Figure 5.6), constructed using the Maple software package.

As can be seen from Figures 5.5 and 5.6, the maximum value of the dynamic uncertainty in measuring the dynamic torque of the rotating system (RS) within the range of 8 to 0.5 N·m at the nominal rotor speed of 50 Hz does not exceed 0.15 N·m (Figure 5.5), while at the minimum rotor speed it is 0.01 N·m. Therefore, the standardized value of dynamic uncertainty should be taken as the maximum value observed at the upper measurement limit of the dynamic torque at the maximum rotational frequency of the rotor system.

From the obtained results of the dynamic uncertainty assessment during dynamic torque measurement, it can be concluded that the relative dynamic uncertainty does not exceed 1.9% within the measurement range of 8 to 0.5 N·m.

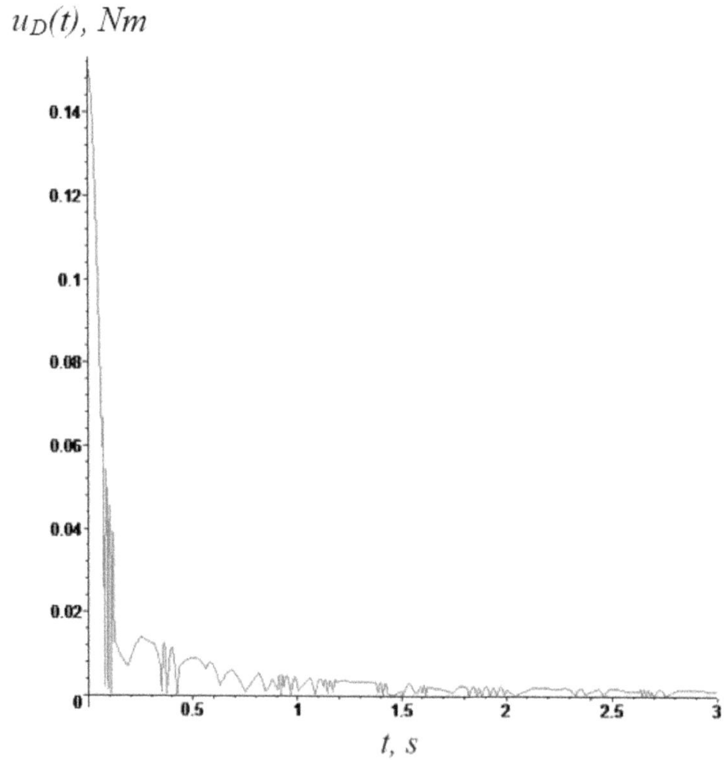

FIGURE 5.5 Dynamic torque measurement uncertainty variation of the RS at the nominal input signal frequency of 50 Hz.

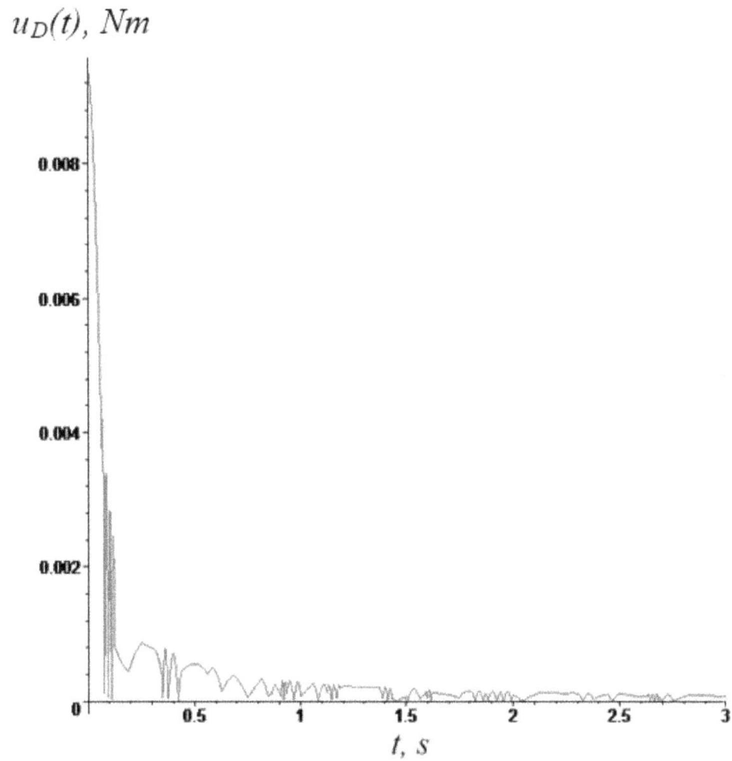

FIGURE 5.6 Dynamic torque measurement uncertainty variation of the RS when the input signal frequency decreases to 1 Hz.

To investigate the characteristics of the dynamic uncertainty in measuring the dynamic torque of the RS within the range from 30 to 1 N·m ($M_k = 30\,\mathrm{N\cdot m}$), the dependencies of the dynamic uncertainties were constructed using the Maple software package. The results for the nominal rotor speed of 50 Hz are shown in Figure 5.7, and for the minimum rotor speed of 1 Hz in Figure 5.8.

As seen in Figures 5.7 and 5.8, the maximum value of the dynamic uncertainty in measuring the dynamic torque within the range of 30 to 1 N·m at the nominal rotor speed of 50 Hz does not exceed 0.62 N·m (Figure 5.7), while at the minimum rotor speed the maximum dynamic uncertainty does not exceed 0.02 N·m. The standardized value of the dynamic uncertainty for measuring the dynamic torque in the range from 30 to 1 N·m should be taken as the maximum value (0.62 N·m), which occurs at the upper measurement limit of the dynamic torque at the initial (maximum) rotor speed.

The results obtained for the assessment of the dynamic uncertainty in measuring the dynamic torque within the measurement range of 30 to 1 N·m indicate that the relative dynamic uncertainty does not exceed 2.1%.

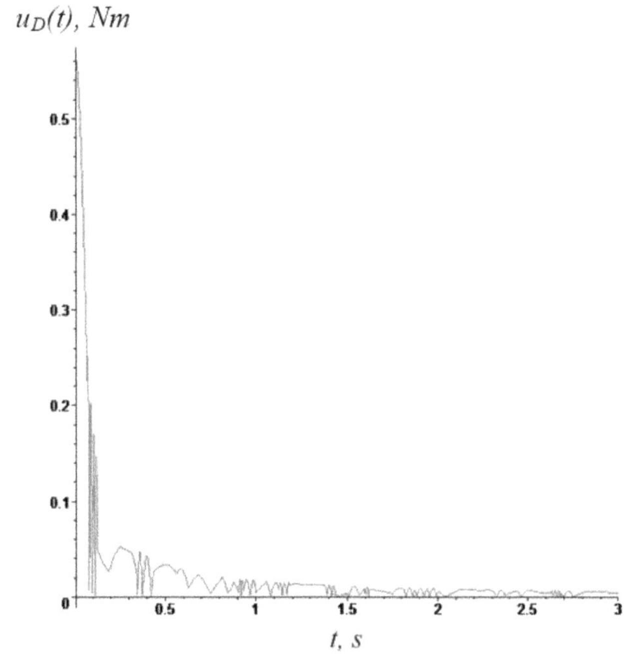

FIGURE 5.7 Characteristic of the dynamic measurement uncertainty of the RS torque for the range of 30 to 1 N·m at the nominal rotor speed of 50 Hz.

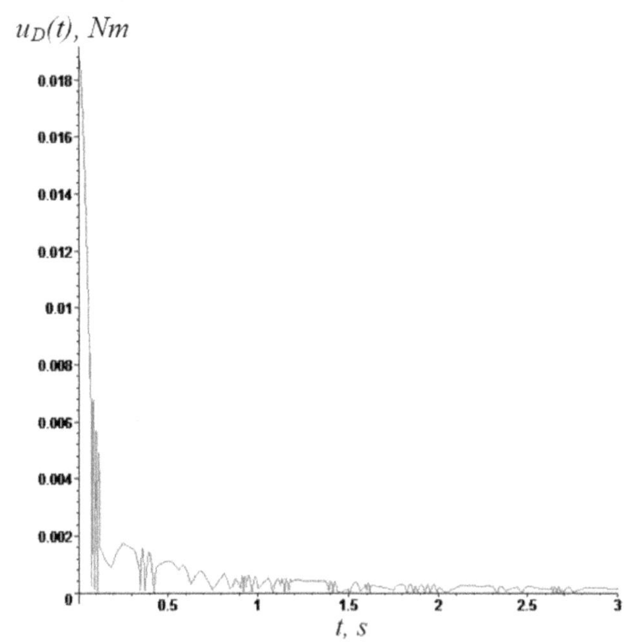

FIGURE 5.8 Characteristic of the dynamic measurement uncertainty of the RS torque for the range of 30 to 1 N·m at the minimum rotational speed of 1 Hz.

5.5 UNCERTAINTY IN SIGNAL RECONSTRUCTION DURING DYNAMIC MEASUREMENTS

When measuring rapidly changing processes, it is essential to account for and correct the dynamic uncertainty of the MI. This uncertainty arises from the non-ideal dynamic properties of the MI and is associated with the non-zero response time of the instrument to the input signal. One of the primary objectives of measurement and instrumentation technology is also to determine the measured quantity from the output signal of the MI $y(t)$.

Correcting the dynamic error of the input signal $x(t)$ requires solving the operator-level equation of the system dynamics:

$$x(t) = B^{-1}[y(t)], \tag{5.52}$$

where $y(t)$ is the output signal of the measuring transducer, and B^{-1} is the inverse operator of the measuring transducer.[11]

It is assumed that estimates of the operator B, the output signal $y(t)$, and their associated uncertainties $u(B)$ and $u[y(t)]$ are known. The solution is understood as determining an estimate of the input signal $x(t)$ and its uncertainty $u[x(t)]$. The uncertainty $u(B)$ represents either the uncertainty in identifying the operator of a specific MI or the deviation of the specific operator from its nominal value. The uncertainty $u[y(t)]$ reflects the uncertainty in determining $y(t)$, which generally consists of the uncertainties of registration and counting. The uncertainty $u[x(t)]$ represents the uncertainty associated with the correction, that is, the uncertainty in determining the dynamic uncertainty. Obviously, correction is meaningful only when $u[x(t)] < u[y(t)]/k$.

Unlike static corrections, where adjustments are introduced directly into the measurement result, compensation of dynamic uncertainty can only be achieved through advanced processing of the output signal of the MI.

Reconstruction of the input signal involves:

1. obtaining the output signal $y(t)$ of the MI with duration T;
2. decomposing the measured signal $y(t)$ into a Fourier series with the period of the first harmonic in the form:

$$y(t) = \frac{m_0}{2} + \sum\nolimits_{k=1}^{K} m_k \cos(k\omega_0 t) + \sum\nolimits_{k=1}^{K} n_k \sin(k\omega_0 t), \tag{5.53}$$

where $\omega_0 = 2\pi/\mathrm{T}$, T is the duration of the output signal $y(t)$, and m_0, m_k, n_k are coefficients determined by the following formulas:

$$m_0 = \frac{2}{N}\sum\nolimits_{n=0}^{N-1} y_n, \tag{5.54}$$

$$m_k = \frac{2}{N}\sum\nolimits_{n=0}^{N-1} y_n \cos(k\omega_0 t_n), \tag{5.55}$$

$$n_k = \frac{2}{N}\sum\nolimits_{n=0}^{N-1} y_n \sin(k\omega_0 t_n); \tag{5.56}$$

3. experimental determination of the amplitude–frequency $A(\omega)$ and phase–frequency $\varphi(\omega)$ characteristics of the measuring transducer, and the subsequent determination of the real and imaginary components of its complex frequency response:

$$a(\omega) = A(\omega)\cos(\varphi(\omega)), \tag{5.57}$$

$$b(\omega) = A(\omega)\sin(\varphi(\omega)); \tag{5.58}$$

4. calculation of the Fourier coefficients c_k and z_k of the input signal $x(t)$ from the system of equations:

$$\begin{cases} c_k a(k\omega) + z_k b(k\omega) = m_k, \\ z_k a(k\omega) - c_k b(k\omega) = n_k; \end{cases} \tag{5.59}$$

$$m_o = \frac{c_0}{a(0)}; \tag{5.60}$$

5. determination of the input signal $x(t)$ in the form of a Fourier series:

$$\mathrm{x}(t) = \frac{c_0}{2} + \sum_{k=1}^{K} c_k \cos(k\omega t) + \sum_{k=1}^{K} z_k \sin(k\omega t). \tag{5.61}$$

A drawback of this procedure is the computational redundancy. Since the actual output signal $y(t)$ is defined only over a finite interval $[0, T]$, it is sufficient to assume that the function $y(t)$ is either even, in which case it can be expanded in cosines, or odd, in which case it can be expanded in sines. This significantly reduces the computational effort and simplifies the correction algorithm.

In practical measuring transducers, whether digital or analog, the output signal $y(t)$ is provided as a set of discrete samples $y(t_i)$. Therefore, when calculating the Fourier series coefficients, the integral sign is replaced by a summation sign. Assuming the function is even, it can be expanded in cosines, limited to the Kth harmonic:

$$y(t_i) = \frac{m_0}{2} + \sum_{k=1}^{K} m_k \cos(k\omega t). \tag{5.62}$$

The coefficients m_0 and m_k are calculated using the following formulas:

$$m_0 = \frac{2}{N}\sum_{i=1}^{N} [y(t_i)T_a], \tag{5.63}$$

$$m_k = \frac{2}{N}\cos(k\omega t)\sum_{i=1}^{N} [y(t_i)T_a], \tag{5.64}$$

where $N = \mathrm{T}/T_a$, and T_a is the signal sampling interval.

The input signal $x(t)$, expanded in a Fourier series using cosines, has a form analogous to Equation (5.62):

$$x(t_i) = \frac{c_0}{2} + \sum\nolimits_{k=1}^{K} c_k \cos(k\omega t), \tag{5.65}$$

where $c_0 = \frac{2}{N}\sum\nolimits_{i=1}^{N}[x(t_i)T_a]$ and $c_k = \frac{2}{N}\cos(k\omega t)\sum\nolimits_{i=1}^{N}[x(t_i)T_a]$.

The coefficients c_0 and c_k are determined using the following equations:

$$c_0 = \frac{m_0}{a(0)}, \tag{5.66}$$

$$c_k = \frac{m_k a(k\omega)}{a^2(k\omega) + b^2(k\omega)}. \tag{5.67}$$

For most measuring transducers $\varphi(\omega) = 0$, so Equation (5.67) can therefore be simplified to the following form:

$$c_k = \frac{m_k}{a(k\omega)} = \frac{m_k}{A(k\omega)}. \tag{5.68}$$

Thus, the reconstructed signal can be described by the following expression:

$$x(t_i) = \frac{T_a\sum\nolimits_{i=1}^{N} y(t_i)}{\mathrm{A}(0)} + \sum\nolimits_{k=1}^{K} \frac{2T_a\sum\nolimits_{i=1}^{N}[y(t_i)]}{\mathrm{A}(k\omega)}\cos^2(k\omega t). \tag{5.69}$$

When reconstructing a signal, its uncertainty must be evaluated. For this task, there are two types of uncertainty: methodological, associated with replacing the infinite summation of signal harmonics with a finite number K, and instrumental, caused by the uncertainty in registering the output signal $y(t)$ and the frequency characteristics of the MI.

Let us consider the instrumental uncertainty in reconstructing the input signal. Measurement of the input signal of the MI is indirect; that is, the quantity being determined, $x(t)$ is inferred from direct measurements of the MI output signal $y(t)$ and the AFR $A(\omega)$ of the MI, to which it is functionally related. Typically, the quantities $y(t)$ and $A(\omega)$ are correlated, since both are determined using the same MI.

Therefore, the algorithm for expressing the uncertainty of signal reconstruction in dynamic measurements is as follows:

1. determine the estimate of the input signal using Equation (5.69);
2. determine the standard uncertainties $u[y(t)]$ and $u[A(\omega)]$. If the measurements of $y(t)$ and $A(\omega)$ are repeated, the uncertainty is evaluated as type A; otherwise, for single measurements, it is evaluated as type B. In practice, measurements of $y(t)$ are often performed once, while the AFR of the MI is determined in advance during production (or through metrological studies);

3. calculate the sensitivity coefficients:

$$\frac{\partial x(t_i)}{\partial y(t_i)} = \frac{NT_a}{A(0)} + \sum_{k=1}^{K} \frac{2NT_a}{A(k\omega)} \cos^2(k\omega t), \tag{5.70}$$

$$\frac{\partial x(t_i)}{\partial A(\omega)} = \sum_{k=1}^{K} \frac{2T_a \sum_{k=1}^{K} [y(t_i)]}{A^2(k\omega)} \cos^2(k\omega t); \tag{5.71}$$

4. determine the correlation coefficient $r[y(t), A(\omega)]$;
5. calculate the squared combined uncertainty of the measurement result using the following equation:

$$u_c^2 = \left(\frac{\partial x(t_i)}{\partial y(t_i)}\right)^2 u^2[y(t)] + \left(\frac{\partial x(t_i)}{\partial A(\omega)}\right)^2 u^2[A(\omega)] +$$

$$+ 2\frac{\partial x(t_i)}{\partial y(t_i)} \frac{\partial x(t_i)}{\partial A(\omega)} u[y(t)]u[A(\omega)]r[y(t), A(\omega)]; \tag{5.72}$$

6. determine the expanded uncertainty:

$$U = t_p(v)u_c. \tag{5.73}$$

Thus, when reconstructing the measured signal, the following sequence of actions is recommended: obtain the output samples of the MI with known uncertainty $u[y(t)]$; experimentally determine the MI AFR and estimate its uncertainty $u[A(\omega)]$; calculate the input signal of the MI using Equation (5.69), specifying the desired level of methodological uncertainty by limiting the number of harmonics to K; and, finally, compute the instrumental uncertainty of the signal reconstruction using Equations (5.70)–(5.73).

Notes

1 S. Eichstädt, A. Link, C. Elster, Dynamic uncertainty for compensated second-order systems, *Sensors*, Volume 10, Issue 8, 2010, 7621–7631.

2 O. M. Vasilevskyi, Evaluation of the uncertainty of output signals of measuring instruments in dynamic operating modes, *Information Processing Systems*, Volume 4, Issue 85, 2010, 81–84.

3 S. Eichstädt, C. Elster, I. Smith, T. Esward, Evaluation of dynamic measurement uncertainty – an open-source software package to bridge theory and practice, *Journal of Sensors and Sensor Systems*, Volume 6, 2017, 97–105.

4 O. Vasilevskyi, V. Kucheruk, I. Kurytnik, An approach to the evaluation of dynamic uncertainty in measurement using non-statistical techniques, *Measurements, Automation, Control*, Volume 60, Issue 11, 2014, 997–1001.

5 O. M. Vasilevskyi, V. M. Didych, *The method of expressing the uncertainty of dynamic measurements, Modern engineering research: topical problems, challenges and modernity: collective monograph*, Riga: Izdevnieciba "Baltija Publishing", 2020, pp. 63–83.

6 O. M. Vasilevskyi, A frequency method for dynamic uncertainty evaluation of measurement during modes of dynamic operation, *International Journal of Metrology and Quality Engineering*, Volume 6, Issue 2, 2015, 202.
7 O. M. Vasilevskyi, M. Yu. Yakovlev, P. I. Kulakov, Spectral method to evaluate the uncertainty of dynamic measurements, *Technical Electrodynamics*, Volume 4, 2017, 72–78.
8 O. M. Vasilevskyi, P. I. Kulakov, K. V. Ovchynnykov, V. M. Didych, Evaluation of dynamic measurement uncertainty in the time domain in the application to high speed rotating machinery, *International Journal of Metrology and Quality Engineering*, Volume 8, Issue 25, 2017.
9 W. Wójcik, O. Vasilevskyi, P. Kulakov, D. Kompanets, O. Lysenko, V. Prysyazhnyuk, D. Baitussupov, A new approach to assessing the dynamic uncertainty of measuring devices, *Photonics Applications in Astronomy, Communications, Industry, and High-Energy Physics Experiments* 2018, 108082E, 2018, 135.
10 O. M. Vasilevskyi, Method of dynamic expression of uncertainty of measurement, *Devices and Methods of Measurements*, Volume 2, Issue 7, 2013, 109–113.
11 O. M. Vasilevskyi, *Estimation of dynamic measurements uncertainty*, Visnyk of Vinnytsia Polytechnical Institute, 3, 2011. pp. 9–13.

BIBLIOGRAPHY

Eichstädt, S., C. Elster, I. Smith, and T. Esward, Evaluation of dynamic measurement uncertainty – An open-source software package to bridge theory and practice, *Journal of Sensors and Sensor Systems*, 6, 2017, 97–105, https://doi.org/10.5194/jsss-6-97-2017

Eichstädt, S., A. Link, and C. Elster, Dynamic uncertainty for compensated second-order systems. *Sensors*, 10, 8, 2010, 7621–7631. https://doi.org/10.3390/s100807621

Vasilevskyi, O., V. Kucheruk, and I. Kurytnik, An approach to the evaluation of dynamic uncertainty in measurement using non-statistical techniques, *Measurements, Automation, Control*, 60, 11, 2014, 997–1001.

Vasilevskyi, O. M., Evaluation of the uncertainty of output signals of measuring instruments in dynamic operating modes, *Information Processing Systems*, 4, 85, 2010, 81–84.

Vasilevskyi, O. M., A frequency method for dynamic uncertainty evaluation of measurement during modes of dynamic operation, *International Journal of Metrology and Quality Engineering*, 6, 2, 2015, 202, https://doi.org/10.1051/ijmqe/2015008

Vasilevskyi, O. M., P. I. Kulakov, K. V. Ovchynnykov, and V. M. Didych, Evaluation of dynamic measurement uncertainty in the time domain in the application to high speed rotating machinery, *International Journal of Metrology and Quality Engineering*, 8, 25, 2017a. https://doi.org/10.1051/ijmqe/2017019

Vasilevskyi, O. M., Estimation of dynamic measurements uncertainty, *Visnyk of Vinnytsia Polytechnical Institute*, 3, 2011. 9–13. https://visnyk.vntu.edu.ua/index.php/visnyk/issue/view/58

Vasilevskyi, O. M., Method of dynamic expression of uncertainty of measurement, *Devices and Methods of Measurements*, 2, 7, 2013, 109–113, https://rep.bntu.by/handle/data/7996

Vasilevskyi, O. M. and V. M. Didych, *The method of expressing the uncertainty of dynamic measurements, modern engineering research: Topical problems, challenges and modernity: Collective monograph*, Riga: Izdevnieciba "Baltija Publishing" (2020): 63–83, ISBN 978-9934-588-47-1, https://doi.org/10.30525/978-9934-588-47-1.4

Vasilevskyi, O. M., M. Y. Yakovlev, and P. I. Kulakov, Spectral method to evaluate the uncertainty of dynamic measurements, *Technical Electrodynamics*, 4, 2017b, 72–78, https://doi.org/10.15407/techned2017.04.072

Wójcik, W., O. Vasilevskyi, P. Kulakov, D. Kompanets, O. Lysenko, V. Prysyazhnyuk, and D. Baitussupov, A new approach to assessing the dynamic uncertainty of measuring devices, *Photonics applications in astronomy, communications, industry, and high-energy physics experiments 2018*, 108082E, 2018, 135. https://doi.org/10.1117/12.2501578

6 Evaluation of Trueness and Precision of Laboratory Test Results

6.1 ON THE DEFINITIONS OF THE CONCEPTS OF ACCURACY CHARACTERISTICS

Laboratory testing (investigation of characteristics) is integral to product quality assurance. The simultaneous presence of both manufacturer and consumer interests raises the issue of ensuring the required accuracy of test results and creating conditions that guarantee their comparability and mutual recognition regardless of the time and place of testing.

The basis for comparing test results is their uniformity, following the principle proclaimed by the International Organization for Standardization (ISO)—"1 + 1 + 1," meaning "one standard – one test method – one conformity assessment procedure."

Testing is a multi-stage procedure that begins with sample preparation and ends with the measurement of a physical quantity carrying information about the characteristic properties of the test object. Each stage of the testing procedure is associated with measurement uncertainty.

The state vector of the observed object $\tilde{Y}$, which corresponds to the actual values of the condition parameters (that, in the general case, differ from the normalized values), is subsequently used as the basis for evaluating the test results.

A non-quantitative assessment of an object's property (*qualitative result*) R_1—for example, a change in shape or color—is obtained through expert judgment or organoleptic methods, based on the object's response to the test conditions as well as to uncontrolled external factors.

In this case, the uncertainty of the non-quantitative assessment is described by the following expression:

$$u(R_1) = F_1(\Delta_1 * \Delta_2 * \Delta_3), \tag{6.1}$$

where F_1 is the symbol of the functional operator, Δ_1 is the uncertainty associated with the specification of the object's operating mode, Δ_2 is the uncertainty associated with the specification of the parameters of the object's operating conditions, Δ_3 is the uncertainty caused by uncontrolled and unmanageable influences, and $*$ is the symbol denoting combination.

If the changes in the state of the object are assessed using quantitative results, then alongside the measurement accuracy of (or specification of) the test conditions, a measurement task arises to evaluate the values of the object's output parameter.

DOI: 10.1201/9781003742708-6

In this case, it is essential to have a correct representation of the object state vector $\tilde{Y}$, considering that, in most cases, the type of relationship between the test conditions and the measured parameter of the object is unknown, which introduces an uncertainty Δ_m.

A quantitative assessment of the test result R_2 is obtained based on the value of the test object parameter $\tilde{y}$ using a measuring instrument (MI), which introduces an additional uncertainty Δ_4. In this case, the uncertainty of the quantitative assessment R_2 can be described by the following expression:

$$u(R_2) = F_2(\Delta_1 * \Delta_2 * \Delta_3 * \Delta_4 * \Delta_m). \qquad (6.2)$$

During control testing, the assessment of the parameter $\tilde{y}$ using a control instrument (CI), which introduces an uncertainty Δ_8, is compared with the established parameter standards $[\hat{y}_{min}, \hat{y}_{max}]$, which have been defined with an associated uncertainty Δ_5. The formation of the parameter standards is performed by converting the calculated normative values into a physical equivalent that is consistent with the obtained assessment of the state vector parameter. The calculation of the standards is carried out with an uncertainty Δ_6, based on input data that have an associated uncertainty Δ_7. The result R_3 of the control testing of the quality indicator of the test object will be known with a combined uncertainty:

$$u_c(R_3) = F_3(\Delta_1 * \Delta_2 * \Delta_3 * \Delta_4 * \Delta_m * \Delta_5 * \Delta_6 * \Delta_7 * \Delta_8), \qquad (6.3)$$

which will influence the formation of the probability of the decision: "the test object is suitable" (S) or "the test object is not suitable" ($\bar{S}$). These decisions will be made with some degree of error, characterized by the producer's risk and the consumer's risk.

Sources of uncertainty can be grouped into two main categories:

1. Random effects, which occur irregularly and with intensities that are difficult to predict, causing dispersion of test results. Examples include uncontrolled and unmanageable influences.
2. Systematic effects, which change in a predictable manner during the testing process and cause bias in the test results, such as the methodology implementing the test procedure. Some sources may simultaneously contribute to both dispersion and bias. Therefore, it is essential to assess their combined influence.

During the attestation or validation of a method, it is necessary to evaluate and then normalize its accuracy indicators, considering possible influences of the testing conditions, characteristic features of laboratory experiment organization, and multi-stage sample preparation. Direct analytical determination of these indicators is practically impossible, as there is usually no mathematical model that accounts for the relationship between the properties of the test object, the conditions and modes of its operation, and the measured output quantity. Consequently, this task is addressed through interlaboratory collaborative experiments, which serve as a physical model for the implementation of the testing procedure.

Competent testing laboratories in a given field are involved in such collaborative experiments, forming a laboratory consortium. This consortium ensures the uniformity of laboratory testing, with its main characteristic being the achievable reproducibility of results at the present time. This approach allows the evaluation and normalization of accuracy indicators to move away from the ideology of setting arbitrary allowable limits; instead, normalization is based on statistical evaluation and reflects the current status of the testing process in laboratories.

From this framework, it follows that the metrological characteristics of MIs included in the test equipment depend on the interrelationship between the parameters of actual test conditions. Therefore, it is necessary to perform experimental evaluation of the metrological characteristics of MIs under realistic testing conditions, where parameter interactions influence the measurement outcome. At the same time, results obtained in one laboratory—with its own organization of the testing procedure and specific conditions—must be compared with results obtained in other laboratories.

Thus, in laboratory testing, the comparability of results is the primary indicator of quality. It is crucial to consider the actual capabilities of laboratories and the organization of testing when implementing attested or standardized methods. The modern approach does not assume that "any laboratory performing measurements according to a given method must achieve an error (uncertainty) not exceeding the limit prescribed by the method." Accuracy requirements can be raised indefinitely, but this would not objectively reflect the technical and organizational capabilities of the laboratory. Therefore, when normalizing test accuracy indicators, it is necessary to establish and then normalize them based on a statistical approach.

Accordingly, the requirements for accuracy indicators of test results and the methods of their evaluation are formulated so that, on one hand, these indicators ensure reliable assessment, and on the other hand, they are presented in a simple and convenient form.

As characteristics of the properties of the measurement (or testing) object, accuracy indicators of results recommended by the international standard ISO 5725-1:2023[1] are used.

The definition of these indicators is a necessary condition for the validation of measurement (testing) methods. Validation is a documented procedure that provides a high degree of confidence that a particular method will yield results conforming to predefined criteria of compliance. During validation, the suitability of the method to ensure reproducibility and trueness of results is established, with accuracy meeting specified requirements.

Validation is required for:

- A new method developed to solve a specific task.
- A recognized method that has been revised, improved, or extended to cover a new task.
- Situations where quality control indicates changes over time in a recognized method.
- A recognized method used in another laboratory, by different operators, or with different instruments or equipment.
- Demonstrating the equivalence of two methods, for example, a new and an existing method.

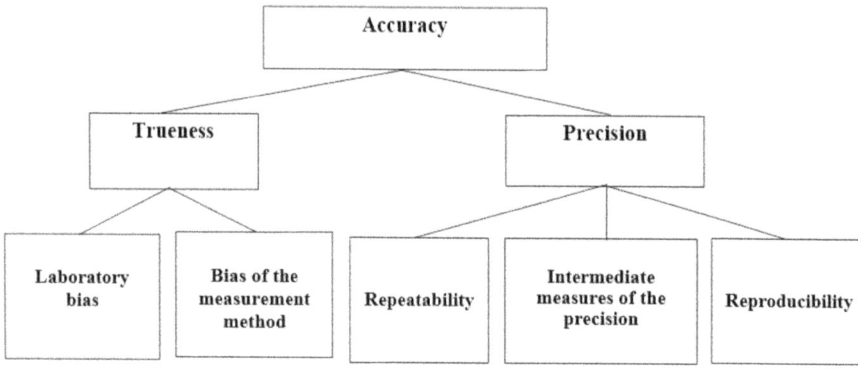

FIGURE 6.1 On the definition of the concept of "accuracy."

The primary qualitative indicator for any experimental procedure is the accuracy of the obtained quantitative result (see Figure 6.1).

According to ISO 5725-1:2023, accuracy is the degree of closeness of the measurement result to the accepted reference (true) value. The accepted reference value is used as the true value (consensus value for comparison) when primary standards are not available. Depending on the task and laboratory capabilities, the accepted reference value may be:

1. A theoretical or established value, based on scientific principles (e.g., molar mass).
2. An assigned or certified value, based on experimental data from national or international organizations (where a reference laboratory exists).
3. A consensus or attested value, based on collaborative work (interlaboratory study or international laboratory consortium).
4. The mathematical expectation of the measured characteristic (mean of observation results) only if the first three options are unavailable.

In this context, accuracy acts as an integral qualitative indicator, characterized by both trueness and precision. The components of accuracy are illustrated in Figure 6.1.

Trueness is characterized by the laboratory bias and the method bias, while the concept of precision is differentiated according to the conditions of the experiment and is referred to as repeatability, intermediate precision, or reproducibility of the result (or method), accordingly.[2]

The trueness of test results reflects the degree of closeness of the mean value to the reference value and is influenced by imperfections in the method, the establishment or control of testing conditions, and the design features of MIs and test equipment. The measure of trueness is the systematic error, and the evaluation of systematic error is expressed as bias.

Thus, bias is the difference between the mathematical expectation of the test results and the accepted reference value. Trueness can only be assessed when information about the accepted reference value of the measured quantity is available, for example, reference materials, certified standard solutions, or standard concentrations.

Components of trueness include:

- Laboratory bias, which characterizes the individual features of the testing procedure organization in a laboratory.
- Method bias, which characterizes the intrinsic trueness of the method.

The evaluation of laboratory bias can be performed in two ways:

- Using reference materials with certified characteristics.
- Comparing results with those obtained by a reference (primary) laboratory.

Since testing is a complex experimental procedure that involves extracting informative parameters or quantities characterizing the properties of the object, the trueness of the method depends on the quality of this extraction process. For example, in chemical analysis, bias may occur if the applied method is unable to fully extract a component, or if the presence of one component interferes with the determination of another component with similar properties.

Precision. Obtaining a test result requires implementing a measurement procedure that is influenced by random variables, which leads to the dispersion of observation results around a certain mean. Many factors can contribute to this dispersion. Among them, five main factors can be identified:

- Operator.
- Equipment used.
- Calibration.
- Environment.
- Time elapsed between measurements.

The characteristic of the dispersion of measurement results, or in other words their stability, is precision—the second, so-called experimental component of accuracy.

Precision describes the closeness of independent test results obtained under specified conditions, and its indicator is the standard deviation of test results performed under these conditions. Both single observations and the mean values of multiple observations can be considered as results. Often, the relative standard deviation is used as a measure of precision; being dimensionless, it is more illustrative.

Precision depends only on random factors (their possible values) and is not related to the reference value or the accepted value. Special attention is given to the test conditions, as they directly influence the variability of results.

Repeatability. As is known from mathematical statistics,[3] the stability of measurement results is reflected in their repeatability under unchanged test conditions. To ensure repeatability, all five factors listed above must remain constant during the experiment. If tests are performed in the same laboratory on identical test objects, by the same operator, using the same setup, and over a short time interval relative to environmental changes, discrepancies between observations are determined solely by the influence of random variables. The dispersion of results within a single laboratory under repeatability conditions is characterized by the within-laboratory variance:

$$Var(e_w) = \sigma_w^2. \tag{6.4}$$

The overall mean of the within-laboratory variances is described by the expression:

$$\sigma_r^2 = \overline{Var(e_w)} = \overline{\sigma_w^2}. \tag{6.5}$$

The overall mean of the within-laboratory variances, obtained during an inter-laboratory study, is normalized as the repeatability variance, which is subsequently adopted as a characteristic of the random component of laboratory error e° for the given method.

In most cases, these discrepancies are well approximated by the normal distribution. The characteristic describing the variability of observation results within a laboratory is repeatability, that is, precision under repeatability conditions. As noted earlier, repeatability conditions require the constancy of the aforementioned factors.

According to the definition of repeatability conditions, measurements intended to determine the repeatability indicator must be performed under stable conditions; that is, during the measurement period, influential factors must remain constant. Specifically, equipment should not be recalibrated between measurements, and testing should be carried out over the shortest possible time intervals to minimize variations in factors such as environmental conditions and the values of standard materials, the stability of which cannot be guaranteed (particularly in biomedical studies). Nevertheless, even under these conditions, results will generally differ within established limits.

The indicator of repeatability is the standard deviation σ_r of a large number of test results obtained under repeatability conditions. A statistical measure of repeatability is the repeatability limit r defined as the value that, with a 95% confidence level, is not exceeded by the absolute difference between results of two measurements (tests) obtained under repeatability conditions.

In some documents, this characteristic is referred to as convergence. However, the term "convergence" is a misinterpretation, and its use in this context is incorrect. In probability theory and mathematical statistics, convergence refers to the property of a sequence of random variables approaching a constant value in probability as $n \to \infty$.

If the tests are conducted in a single laboratory only, the concepts of repeatability and precision are considered identical.

Reproducibility. Tests of the same object, performed using the same method, may be carried out by multiple laboratories under different—but specified or normalized—conditions, equipped with instruments having different (individual) technical characteristics, with experiments conducted by different operators at different times.

Differences in laboratory equipment characteristics, operator qualifications, and the organization of testing procedures within laboratories—even when these remain within permissible limits—ultimately lead to differences in the laboratory components of bias $B°$, the variability of which is characterized by the interlaboratory variance $Var(B°) = \sigma_L^2$. If necessary, the influence of these components can be evaluated separately under intermediate precision conditions.

Since the laboratory error contains a random component, and its systematic component represents one of the possible values of the random variable "interlaboratory variation $B°$," the reproducibility variance σ_R^2 must account for both the repeatability variance and the interlaboratory variance:

$$\sigma_R^2 = Var(B°) + \overline{Var(e_w)}, \tag{6.6}$$

that is

$$\sigma_R^2 = \sigma_L^2 + \sigma_r^2. \tag{6.7}$$

Results obtained in this way (under repeatability conditions) may differ. This property of repeated tests is characterized by a specific indicator—reproducibility of test results, that is, precision under reproducibility conditions. At the same time, repeatability conditions must be maintained within each laboratory.

The indicator of reproducibility is the standard deviation σ_R of measurement (test) results obtained under reproducibility conditions, following the prescribed test method and stable characteristics of the test object.

The reproducibility variance forms the basis for the comparability and verification of conformity of results obtained across laboratories, as well as for method certification, standardization, and validation. It is also an important characteristic in the accreditation of laboratories.

Repeatability and reproducibility represent two extreme cases of precision: the former reflects the minimum variability, and the latter—the maximum possible variability of results.

Under reproducibility conditions, it is assumed that measurement (test) results are obtained using the same method on identical test objects, but in different laboratories, by different operators, and using their own instruments of the same type. The reproducibility limit R is defined as the value which, for the set of all possible pairs of measurement results obtained under reproducibility conditions, is not exceeded by the absolute difference between two measurements within each pair, with a 95% confidence level. The introduction of these limits ensures compatibility (acceptability) of the obtained results.

Thus, the repeatability limit is defined as

$$r = 1.96\sqrt{2\sigma_r^2} = 2.77\sigma_r, \tag{6.8}$$

and the reproducibility limit is defined as

$$R = 1.96\sqrt{2\sigma_R^2} = 2.77\sigma_R. \tag{6.9}$$

For practical use, the coefficient is conventionally rounded to 2.8.

Intermediate precision. Between the conditions of repeatability and reproducibility specified by a method, there are numerous practical cases in testing where some of the assumed conditions are not fully met. An example is a laboratory operating in two shifts, where experiments are conducted by different operators at different times. To characterize such situations, the concept of *intermediate precision conditions* is introduced. The number of types of intermediate precision corresponds to all possible combinations of four elements: time, calibration, operator, and equipment. The measure of intermediate precision is the standard deviation σ_I of test results obtained experimentally under these conditions. When reporting these values, it is recommended to indicate which factors vary: time T, calibration C, operator O, and equipment E ($\sigma_I(T)$, $\sigma_I(C)$, $\sigma_I(O)$, $\sigma_I(E)$). The use of some types of intermediate precision may be applied, for example, in comparative evaluation of laboratories performing tests according to the same method, in method validation within a laboratory, as well as in quality control of laboratory work and operator competence assessment.

Among all types of precision, reproducibility holds special significance. Testing, in addition to measuring the output quantity of an object, involves a combination of interdependent procedures such as sample component extraction, perception of informative quantities characterizing the object's properties, and assignment of conditions and operational modes, which may only be implemented within certain limits, leading to non-unique outcomes. The measured quantity is a functional of the actual values occurring during these procedures, yet the exact analytical relationship describing this functional cannot be established. Therefore, experimentally determined reproducibility is a composite indicator of the testing method, reflecting the potential dispersion of results obtained independently of time and place. Reproducibility values allow comparison not only of specific data obtained through the method but also of the capability of laboratories specialized in these tests to implement the method effectively.

When improving laboratory testing procedures or employing more modern equipment, reproducibility may change, serving either as a characteristic of the compatibility of laboratories performing tests on the same product, or as a measure of the effectiveness (or limitations) of the method when executed by different laboratories.

Thus, precision indicators can be represented in three levels:

1. Highest level: the concept of *accuracy*, measured by the overall measurement error (uncertainty).
2. Second level: accuracy broken down into *trueness* and *precision*, evaluated by bias and the standard deviation of precision, respectively.
3. Third level: components of trueness—*laboratory bias* and *method bias*; and components of precision differentiated according to measurement conditions, characterized by standard deviations of *repeatability*, *intermediate precision*, and *reproducibility*.

This approach to defining "accuracy" aligns with the concept of measurement uncertainty,[4,5] which can be extended to testing: trueness is characterized by Type

B combined uncertainty, while precision, evaluated from experimental data representing direct measurements, is characterized by Type A evaluated uncertainty. All influential factors are considered, establishing their cause-and-effect relationships. In cases where such relationships are complex, the measurement uncertainty may be directly assessed from trueness indicators and the components of precision variance under asymptotically normal estimates and sufficiently large samples.

6.2 EXPERIMENTAL APPROACH TO THE EVALUATION OF MEASUREMENT UNCERTAINTY IN TEST RESULTS

The model-based approach for the evaluation of measurement uncertainty allows estimating the contributions of individual uncertainty components and, if necessary, reducing their impact on measurement accuracy. This approach is particularly suitable, and may be the only feasible method, in cases where statistical observations cannot be conducted, for example, in destructive testing, for economic reasons, or when random errors can be neglected using only prior Type B information. In this approach, the uncertainty evaluation is based on formalizing the influence of all components of the measurement process. Factors involved are described by corresponding input quantities that form a model of indirect measurements, representing the algorithm for transforming these inputs into the output (measured) quantity.

During testing, many sources of uncertainty cannot be formalized in any quantitative manner. These include factors that either cannot be quantified or for which it is difficult to establish a relationship with the measured quantity within the measurement model. Such factors act at all stages of the testing process, including the initial stage, for example, the operator, equipment drift, incomplete specification of the measured quantity, or inadequacy of the measurement model. The influence of most factors characterizing external measurement conditions, such as electromagnetic, vibro-acoustic, radiation, mechanical, and other effects, is difficult to express analytically.

During analytical and biomedical studies, the measurement procedure is preceded by a series of specific steps related to sample selection and preparation, which contribute significantly to the uncertainty of the test results. The impossibility of analytically representing the influence of most external and internal factors on the output measured quantity implies that the effect coefficients for individual factors cannot be determined, especially since this information is often unavailable. The lack of information about the dispersions and covariances of input quantities also arises because some of them have no quantitative expression.

In experimental studies, the measured result is affected by random variables e°, the dispersion of which is characterized by the repeatability variance, representing the average of laboratory dispersions. In addition, each laboratory has its own constant bias component B° because each laboratory conducting the same type of testing has equipment with its "own" distinctive characteristics, but within specified limits, operators with individual characteristics, and its "own" level of test organization within regulated bounds, including procedural control features. To assess possible values of the laboratory bias component B° in an interlaboratory experiment, the standard deviation σ_L is used, characterizing the interlaboratory variability of measurement results.

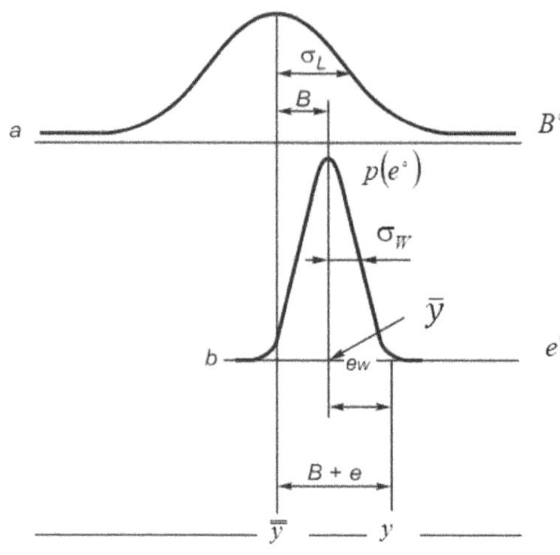

FIGURE 6.2 Components of the combined measurement uncertainty in testing: (a) distribution law of possible values of B°; (b) distribution law of the laboratory random error e°.

The situation described above is illustrated in Figure 6.2, where the laboratory bias component B°, as outlined, represents the center of the distribution of random error e_w in that laboratory.

In evaluating the accuracy characteristics of a test method, the correctness and reproducibility of results obtained according to the method are of primary importance. The reference point is the overall mean $\bar{\bar{y}}$ of results from an interlaboratory study conducted by a set of laboratories. The mathematical expectation of this mean may differ from the reference value μ. Each individual test result Y can be represented statistically by a model, which forms the basis for further analysis:

$$Y = m_{\bar{\bar{y}}} + B^\circ + e^\circ, \tag{6.10}$$

where $m_{\bar{\bar{y}}}$ is the expected value of the overall mean of measurement results, B° is the possible value of the laboratory bias component of the result (correctness) under reproducibility conditions, e° is the random component of measurement error under repeatability conditions, and δ is the bias of the result introduced by the imperfection of the test method.

The application of a statistical model of test results for estimating its uncertainty allows automatic consideration of input factors (including unknown ones), which constitute the primary source of the dispersion of results. The precision obtained in this way corresponds to the combination of factors characterizing specific measurement conditions during testing. This approach, associated with aggregating influencing factors, is discussed in ISO 5725.

In this approach, single measurements of the output quantity of the object are treated as the basic form of measurement. The measurement uncertainty is attributed to individual results, while repeatability, reproducibility, and trueness refer to the execution of the test process according to the method used by all laboratories.

Metrics for evaluating precision, as well as uncertainty, are the standard deviations of measurement results obtained under conditions of repeatability, reproducibility, and intermediate precision. These estimates, obtained in advance from interlaboratory experiments, adequately characterize the "dispersion of values that could reasonably be attributed to the measured quantity," derived directly from direct measurement results.

According to standards, uncertainty should be calculated as the combined uncertainty of all stages of the measurement process. Focusing on input quantities, this approach is called the "bottom-up" (or Type B) approach to uncertainty evaluation in ISO/TS 21748:2004.[6] In other words, it assumes that all uncertainty components can be theoretically or experimentally estimated and summed according to prescribed rules.

As previously noted, this approach is not entirely suitable for testing laboratories, as it assumes that the measurement process model is known and input data are either measured or assigned. In most cases for testing laboratories, the measurement process model is unknown. Therefore, an alternative approach, referred to as the "top-down" approach in ISO/TS 21748:2004, is preferred. It is based on the fact that the performance characteristics of a standard measurement method and the obtained results are independent of the measurement process model and are described by precision and trueness metrics, typically obtained from interlaboratory studies. This assumes that laboratory test procedures are carried out according to a standardized method, making quality control of the test process within the laboratory mandatory.

The fundamental principles for applying this approach are:

- The standard deviation of reproducibility obtained from interlaboratory experiments provides a valid basis for evaluating the uncertainty of measurement results in laboratory tests.
- Influences not observed during interlaboratory experiments must be negligible or accounted for by extending the basic statistical model (6.10) with additional components.

Considering that the standard uncertainty of a test result corresponds to the dispersion of measurements of the output quantity, the expression for the combined uncertainty of laboratory results according to the basic statistical model is given by:

$$u_c(y) = \left[u^2(B^\circ) + u^2(e^\circ)\right]^{0.5}. \qquad (6.11)$$

When the standardized method is fully implemented, this approach, based on measurements of the output quantity, allows for the estimation of the uncertainty of test results for any laboratory.

Considering the limited volume of experimental data within a single laboratory, the interlaboratory variance $\sigma_L{}^2$ is used instead of $u^2(B^\circ)$, and the repeatability variance $\sigma_r{}^2$ is used instead of $u^2(e^\circ)$. Consequently, the equation for estimating the standard uncertainty of a result obtained during testing in a laboratory can be expressed as:

$$u_c(\hat{y}) = \left[\sigma_L^2 + \sigma_r^2\right]^{0.5}. \tag{6.12}$$

The right-hand side of expression (6.12) represents the estimate of the reproducibility variance.

Thus, the estimate of the standard deviation of reproducibility σ_R fully characterizes the uncertainty of the underlying statistical model

$$u_c(\hat{y}) = u_R, \tag{6.13}$$

where $u_R = \sigma_R$.

The standard uncertainty $u\left(\widehat{\delta}\right)$, caused by the method bias, is determined by the expression

$$u\left(\widehat{\delta}\right) = \left[\sigma_{\widehat{\delta}}^2 + u^2(\mu)\right]^{0.5}, \tag{6.14}$$

where $u(\mu)$ is the standard uncertainty corresponding to the possible deviation of the certified value μ, which is used for assessing trueness in the interlaboratory study.

Thus, the standard uncertainty of the standardized method can be expressed as

$$\acute{u}(y) = \left[u_R{}^2 + u^2\left(\widehat{\delta}\right)\right]^{0.5}. \tag{6.15}$$

From the above, it can be concluded that the observed standard deviation of the method bias δ, the laboratory component of the bias B, and the influence of random effects e fully characterize the variability of results under interlaboratory studies and are used in normative documents to regulate the precision indicators of a test method.

If there are grounds to assume that the tests in a laboratory are performed fully in accordance with the method, then for the case when the test result is determined according to the basic statistical model, the normalized value $\acute{u}(y)$ can be used as a reference for evaluating the uncertainty of results obtained in the laboratory.

In practice, u_R and $u(\widehat{\delta})$ do not necessarily include all possible sources of variability affecting measurement results. The absence of an assessment of some important factors present during routine laboratory testing is due to the nature of interlaboratory studies, where all laboratories operate under identical specified conditions, the experimental plan does not provide for them, or their absence occurs simply by chance. All effects observed or identified during the interlaboratory study, if considered significant, should be considered. However, even in this case, it is impossible to account for everything in an interlaboratory experiment.

Preparation of completely homogeneous material samples or test objects in different laboratories, as well as their storage, is challenging. Moreover, each laboratory has its own peculiarities in the organization of testing. In such cases, the standard deviation of repeatability and, especially, the standard deviation of reproducibility obtained from the interlaboratory study will be biased due to differences in the samples used by specific laboratories.

All these deviations can be evaluated by extending the basic statistical model, introducing additional components, that is, by supplementing the uncertainty with contributions calculated using the modeling approach

$$y = \mu + \delta + B + e + \sum_{i=1}^{n} c_i \acute{x}_i, \tag{6.16}$$

where $\acute{x}_i$ are the values deviating from the nominal x_i that were not accounted for in the interlaboratory study, and c_i are the sensitivity coefficients.

Based on the model described by Equation (6.16), the uncertainty can be estimated as follows:

$$u''(y) = \left[\sigma_R{}^2 + u^2\left(\widehat{\delta}\right) + \sum_{i=1}^{n} c_i^2 u^2(\acute{x}_i)\right]^{0.5}, \tag{6.17}$$

where $\sigma_R{}^2$ is the combined contribution of the laboratory bias component B and the random component $eu\left(\widehat{\delta}\right)$ is the uncertainty associated with the method bias δ, determined based on measurements of a reference standard or comparison sample with a certified value μ, and $u(\acute{x}_i)$ are the uncertainties associated with the input quantity $\acute{x}_i$.

In general, Equation (6.17), considering the effects not considered in interlaboratory studies, can be applied to any precision conditions by summing the contributions.

The bottom-up (or "descending") approach accounts for differences between the conditions of measurements in a specific laboratory and the conditions of the interlaboratory experiment. In this approach, the uncertainty estimate is represented by a variance whose components include these differences as additional terms.

Uncertainty values obtained using the top-down and bottom-up approaches will differ when unknown or unanticipated effects are present.

The bottom-up approach should only be applied when the laboratory bias and repeatability components are under control, for example, using Shewhart control charts. When excessive bias is detected, corrective actions must be undertaken to bring the bias within the required range before continuing measurements. Such actions typically involve investigations and elimination of the cause of the bias.

Thus, the bottom-up approach is more advantageous for estimating the uncertainty of laboratory test results when standardized values of repeatability and reproducibility variances are available, with mandatory internal laboratory control. Combining elements of the experimental (bottom-up) approach, which does not require knowledge of cause-and-effect relationships, with elements of the modeling (top-down) approach allows a reliable evaluation of the uncertainty of quantitative test results.

6.3 EXAMPLE OF ESTIMATING THE UNCERTAINTY OF TEST RESULTS FOR DETERMINING THE VOLUME FRACTION OF BENZENE

The laboratory performs tests to determine the volume fraction of benzene in gasoline for a wide range of clients according to a standardized method, which does not provide information on the uncertainty of the results. However, the method specifies:

- Repeatability limit (the maximum permissible difference between results obtained under repeatability conditions), $r = 0.08\%$.
- Reproducibility limit (the maximum permissible difference between two parallel measurements obtained under reproducibility conditions), $R = 0.18\%$.

Additionally, the method notes that "heavy aromatic compounds" interfere with the determination of benzene. Special measures are required to ensure sample stability during storage and analysis.

According to the method, trueness under interlaboratory conditions has not been evaluated. The method allows determination of the benzene volume fraction in the range from 0.1% to 5.0%.

To calculate the uncertainty of the test result, the relationships between the standard deviations of repeatability and reproducibility, as well as the corresponding limits, are used, namely: $r = 2.8\sigma_r$ and $R = 2.8\sigma_R$.

Knowing the numerical values of repeatability r and reproducibility R, the standard deviations of repeatability can be obtained using Equation (6.8) and of reproducibility using Equation (6.9): $\sigma_r = 0.08/2.8 = 0.029\%$ and $\sigma_R = 0.18/2.8 = 0.064\%$.

If the conditions under which routine tests are conducted fully coincided with those of the interlaboratory collaborative experiment, that is, reproducibility encompassed all sources of uncertainty, then for a single measurement result, one could directly proceed to calculate the expanded uncertainty U.

For the extreme points of the range, the test results would be as shown in Table 6.1.

Assume that in each laboratory, 2 parallel measurements were performed, that is, $n = 2$, and the result is calculated as the mean value. Then, the uncertainty of the mean result can be calculated using the expression:

$$u = \left[\sigma_R{}^2 - \sigma_r{}^2\left(1 - \frac{1}{n}\right)\right]^{0.5}. \tag{6.18}$$

As a result, we obtain a standard uncertainty of $u = 0.061\%$ and an expanded uncertainty of $U = 0.12$.

This result indicates that increasing the number of parallel observations does not significantly reduce the measurement uncertainty, since reproducibility is the dominant factor. In this case, parallel observations are carried out primarily to reduce the likelihood of gross errors.

TABLE 6.1
Test Results at the Extreme Points of the Range

Benzene Content	$u=\sigma_R$	$U=1.96u$	Results
0.1%	0.064	0.125	0.1 ± 0.1
5.0%	0.064	0.125	5.0 ± 0.1

TABLE 6.2
Calculation Results of Uncertainties Considering Additional Factors

Benzene Content (%)	u_R	u_{inst}	u_{interf}	u_c	U=1.96u_c	Results
0.1	0.064	0.002	0.01	0.065	0.127	0.1 ± 0.1
5.0	0.064	0.1	0.5	0.514	1.007	5.0 ± 1.0

In the interlaboratory experiment, the effect of sample matrix variation (presence of interfering components) was not considered, since all laboratories tested samples with the same composition. In routine testing, however, samples provided by different clients may vary in composition, which was not considered in the interlaboratory experiment. Let us assume that additional studies provided a value of 0.1 relative to the measured concentration.

During the interlaboratory experiment, reproducibility was assessed using specially prepared samples, so their instability was not considered. In routine laboratory work, the sampling and storage of the sample prior to delivery to the laboratory will introduce additional uncertainty in the result, estimated as 0.02 relative to the measured concentration.

The influence of these components must be considered according to expression (6.17) with a sensitivity coefficient $c_i = 1$.

The combined uncertainty, whose components were determined using the bottom-up and top-down approaches, is calculated using the equation:

$$u_c = \left[u_R^2 + u_{\text{inst}}^2 + u_{\text{interf}}^2\right]^{0.5}. \qquad (6.19)$$

The calculation results of uncertainties are presented in Table 6.2.

For intermediate concentration levels, the uncertainty is calculated using the equation of a straight line passing through two points:

$$U = 0.18X + 0.109. \qquad (6.20)$$

Thus, the successful combination of the two approaches makes it possible to objectively assess the uncertainty of the result.

From the given example, it follows that ignoring additional factors, especially at high concentrations, may lead to an overestimation of the quality of the results and, consequently, to an incorrect conclusion about the quality of the benzene.

6.4 APPLICATION OF ROBUST METHODS IN THE EVALUATION OF TEST RESULT REPEATABILITY

When validating testing methods aimed at establishing and subsequently standardizing accuracy characteristics, a collaborative laboratory study is generally conducted. The assessment of the method's trueness is typically preceded by a precision study. Based on the results of these collaborative tests, the standard deviation (SD) is calculated for each laboratory. The ensemble of these SDs is then used to compute the repeatability standard deviation, which represents the lower bound of the method's precision.

According to standards,[7] in assessing inter-laboratory comparability, laboratory results are examined for homogeneity using Cochran's G-test, which identifies "quasi-outliers" and "statistical outliers." Traditionally, SD values deemed distorted are excluded, and subsequent data processing is performed without them. However, both the exclusion and inclusion of these data can lead to an inadequate representation of the collaborative test results, especially in small sample sizes, where each observation carries significant weight. The uncertainty in SD estimation for small samples ($n < 10$) may vary by tens of percent.

Experimental data are generally assumed to follow a Gaussian (normal) distribution. In practice, particularly for small samples, deviations from normality occur, and classical statistical tests, which are asymptotic in nature, cannot be directly applied. Consequently, the assumed distribution may differ from the actual distribution of the experimental data.

Nevertheless, the classical parametric approach, which assumes normality, has become so entrenched in practice that abandoning it is impractical. These methods are straightforward and intuitive, and a comprehensive statistical theory supports their use. It should be noted that Cochran's criterion is quite sensitive to deviations from normality. Hence, there arose a need to adapt the "old" model to modern challenges, that is, to develop methods capable of accounting for gross errors under certain conditions while still providing reasonably accurate parameter estimates based on available data. These methods are known as robust methods.

In statistics, robustness refers to the insensitivity of an estimator to deviations and inhomogeneities in the sample, which are generally associated with unknown causes.

Robust methods are based not on a simple normal distribution but on a mixture distribution—a normal model contaminated by rare, extreme observations. The central part of the distribution remains stable and corresponds to the probable normal model, while the tails are extended and characterized by relatively sharp outliers (or quasi-outliers for small samples) of individual measurement results. This approach preserves the conventional, intuitive notion of the homogeneity of the hypothetical population, on which probabilistic estimates are based, while introducing the possibility of large deviations in the tails of the distribution. The tails are typically constrained according to the chosen statistical indicator.

Among robust methods and their implementation algorithms, the classical Huber approach is the most widely applied. It uses a parameter k, determined based on the contamination of the statistical population, which defines the modification step for infrequently deviant observations:

$$\widehat{\theta} = \frac{1}{n}\left[\sum\nolimits_{|x_i - \theta| < k} x_i + (n_1 + n_2)k\right], \tag{6.21}$$

where $\widehat{\theta}$ is the robust estimator determined through iterative procedures, k is the tuning constant (allowed deviation from the central location of the population), taken as a fixed value considering the proportion of "gross errors" (outliers) in the data set,

n_1 is the number of observations in the set with values $x_i < (\theta - k)$, and n_2 is the number of observations in the set with values $x_i > (\theta + k)$.

At the initial stage of the calculation, in accordance with (6.21), the median of the sample is typically taken as the estimate $\widehat{\theta}$, since it is more robust to outliers than the arithmetic mean. The so-called "outliers" are not excluded from the available data but are replaced by the corresponding values $x_i - k$ and $x_i + k$. Based on both the "true" (original) and the modified data, an updated estimate is determined at each subsequent jth iteration step, and the iteration continues until all original data fall within the interval of "true" values, that is, $|x_i - \theta| < k$.

The most widely used approach is based on the so-called "S-algorithm," which is aimed at ensuring robustness of the estimate of the variance—for example, the estimate of repeatability.

A necessary condition for applying this method is that the robust estimate S_j^* must be unbiased, that is, at each jth iteration step it is adjusted ("fitted") to the standard deviation σ of the original population. To ensure unbiasedness, a consistency factor ξ is introduced, with the requirement that:

$$E\left[(\xi S^*)^2\right] = \sigma^2. \tag{6.22}$$

The symbol "$*$" will hereinafter denote a robust estimate.

The robust estimate of the standard deviation S^* shall remain within the specified limits of robustness to outliers with a given probability. For this purpose, a restriction is introduced on the maximum deviation $\eta\sigma$ from the center of the distribution, expressed as the inequality:

$$P[S^* > \eta\sigma] = \alpha, \tag{6.23}$$

where σ is the standard deviation of the population, assumed to be normally distributed according to the available experimental data under the assumption of a "pure" normal distribution law, η is a limiting factor that depends on the number of elements in the sample, and $P = (1 - \alpha)$ is the probability that the upper constraint on the permissible deviation of the robust estimate of the standard deviation S^* is satisfied (most often this parameter is taken with $\alpha = 0.1$).

Thus, the limiting factor η is chosen as the upper $(1 - \alpha) \times 100\%$ quantile of the distribution.

Consequently, we arrive at the conclusion that the estimate of the deviation from the central parameter k in expression (6.21) is given by:

$$\psi_j = \eta S_j^*, \tag{6.24}$$

where S_j^* is the robust estimate of the standard deviation calculated at the jth iteration step.

Based on the value of ψ_j obtained at this step, the modified values for the set of considered variances S_{ij}^* are determined, after which an updated robust estimate of the repeatability standard deviation is calculated:

$$S^*_{j+1} = \xi \sqrt{\sum_{i=1}^{n} \frac{\left(S^*_{ij}\right)^2}{n}}, \tag{6.25}$$

where S^*_{ij} is the robust estimate of the standard deviation calculated at the jth iteration step for the ith laboratory participating in the interlaboratory experiment (n is the total number of participating laboratories).

The calculated value S^*_{j+1} is then used to determine a new limited value ψ_{j+1}.

The iterative procedure continues until all initial values of the standard deviations from the participating laboratories fall within the limits of the current threshold value.

The estimate of the variance—and therefore of the standard deviation—depends on the number of degrees of freedom $v = n - 1$.

Taking this into account, by multiplying both sides of expression (6.22) by v, we obtain:

$$E\left[v\left(\frac{S^*}{\sigma}\right)^2\right] = \frac{v}{\xi^2}. \tag{6.26}$$

In the left-hand side of equality (6.26), the statistic χ^2 appears.

Thus, the unbiasedness condition for the robust estimate of the variance can be written as:

$$E\left[\chi^2\right] = \frac{v}{\xi^2}. \tag{6.27}$$

Based on expression (6.23), we can determine an equivalent upper probabilistic bound for the random variable χ^2:

$$P\left[\chi^2 > v\eta^2\right] = \alpha. \tag{6.28}$$

Thus, the expected value of the random variable χ^2 can be represented as the sum of two components:

$$E\left[\chi^2\right] = \int_0^{v\eta^2} x p_\chi(x)dx + \int_{v\eta^2}^{+\infty} x p_\chi(x)dx, \tag{6.29}$$

where $p_\chi(x)$ is the probability density function of the random variable x with v degrees of freedom:

$$p_\chi(x) = e^{\left(-\frac{x}{2}\right)} x^{\left(\frac{v}{2}-1\right)} \frac{2^{\left(-\frac{v}{2}\right)}}{\Gamma\left(\frac{v}{2}\right)}, \tag{6.30}$$

where $\Gamma\left(\frac{v}{2}\right)$ is the Gamma function.

Since we are considering the robust estimate of the standard deviation with an upper bound given by expression (6.23), the second integral on the right-hand side of expression (6.29) can be approximated by a uniformly distributed random variable. In this approximation, it is assumed that the "tail" of the distribution χ^2 containing $\alpha \cdot 100\%$ of the possible values of the given random variable consists of equally probable value $x = \nu\eta^2$:

$$\int_{\nu\eta^2}^{+\infty} x p_{\chi}(x)dx = \nu\eta^2 \int_{\nu\eta^2}^{+\infty} f(x)dx = \alpha\nu\eta^2. \tag{6.31}$$

Based on expression (6.28), the value of ${\chi_{P=1-\alpha}}^2$ is obtained from Pearson distribution tables, and then, using expression (6.31), the limiting factor η is determined, for which condition (6.23) is satisfied. Thus

$$\eta = \frac{{\chi_{P=1-\alpha}}^2}{\nu}. \tag{6.32}$$

Now, let us consider the first term on the right-hand side of expression (6.29), which also implicitly depends on the number of degrees of freedom ν.

We introduce the parameter $\omega = \nu + 2$ and make use of the property of the Gamma function:

$$\Gamma\left(\frac{\omega}{2}\right) = \left(\frac{\omega}{2} - 1\right)\Gamma\left(\frac{\omega}{2} - 1\right). \tag{6.33}$$

Considering that $\frac{\nu}{2} = \left(\frac{\omega}{2} - 1\right)$, we can write:

$$\Gamma\left(\frac{\omega}{2}\right) = \frac{\nu}{2}\Gamma\left(\frac{\nu}{2}\right), \tag{6.34}$$

from which it follows that

$$\Gamma\left(\frac{\nu}{2}\right) = \frac{2}{\nu}\Gamma\left(\frac{\omega}{2}\right). \tag{6.35}$$

Let us substitute relation (6.35) into expression (6.30), and also consider that a simplification can be made in this expression:

$$xx^{\left(\frac{\nu}{2}\right)} = x^{\left(\frac{\omega}{2}-1\right)}. \tag{6.36}$$

Then, the first term on the right-hand side of expression (6.29) can be represented as:

$$\int_0^{\nu\eta^2} xe^{\left(-\frac{x}{2}\right)}x^{\left(\frac{\nu}{2}-1\right)}\frac{2^{\left(-\frac{\nu}{2}\right)}}{\Gamma\left(\frac{\nu}{2}\right)}dx = \nu\int_0^{\nu\eta^2} e^{\left(-\frac{x}{2}\right)}x^{\left(\frac{\omega}{2}-1\right)}\frac{2^{\left(-\frac{\omega}{2}\right)}}{\Gamma\left(\frac{\omega}{2}\right)}dx. \tag{6.37}$$

Under the integral sign in expression (6.37), $p_{\chi,\omega}(x)$ represents the probability density function of the random variable χ^2 with ω degrees of freedom. Based on this, expression (6.37) corresponds to the definition of vz, where the value of $z = P({\chi_\omega}^2 \leq v\eta^2)$ is obtained from the Pearson distribution table.

As a result of the analysis, expression (6.31) can be written in the form of the equality:

$$E\left[\chi^2\right] = vz + \alpha v\eta^2. \tag{6.38}$$

On the other hand, based on expression (6.26) for the expected value, we have:

$$E\left[v(S^*\sigma)^2\right] = \frac{v}{\xi^2}. \tag{6.39}$$

It follows from this that

$$\xi = \frac{1}{\sqrt{z + \alpha\eta^2}}. \tag{6.40}$$

Expression (6.40) allows for the calculation of the consistency factor ξ, which ensures that the robust estimate of the standard deviation is unbiased for a given choice of the limiting factor η.

Table 6.3 presents the number of degrees of freedom v and the corresponding values of the limiting factor η and the consistency factor ξ.

TABLE 6.3
Numerical Values of Degrees of Freedom and the Corresponding Values of the Limiting Factor η and the Consistency Factor ξ

Degrees of Freedom, v	Limiting Factor, η	Consistency Factor, ξ
1	1.645	1.097
2	1.517	1.054
3	1.444	1.039
4	1.395	1.032
5	1.359	1.027
6	1.332	1.024
7	1.310	1.021
8	1.292	1.019
9	1.277	1.018
10	1.264	1.017

6.5 EXAMPLE OF APPLYING ROBUST METHODS FOR THE ASSESSMENT OF REPEATABILITY IN LABORATORY TESTS

For the conduct of an interlaboratory collaborative experiment $n = 9$ laboratories with proven experience in this type of testing were involved. On homogeneous test objects, each laboratory performed two trials, based on which the absolute differences were calculated:

$$w_i = |x_{i1} - x_{i2}|, i = \overline{1, n}, \tag{6.41}$$

where x_{i1} and x_{i2} are the results of the experiment in the ith laboratory.

The calculated values of the differences are as follows:

$$w_1 = 0.28,\quad w_2 = 0.49,\quad w_3 = 0.40,\quad w_4 = 0.00,\quad w_5 = 0.35,\quad w_6 = 1.98,$$
$$w_7 = 0.80,\quad w_8 = 0.32,\ w_9 = 0.95.$$

Considering the relation for a sample of two elements, we write:

$$S_i^2 = \frac{|x_{i1} - x_{i2}|^2}{2}. \tag{6.42}$$

Next, for the assessment of repeatability, we will consider the quantity $\sum_{i=1}^{n} w_i^2$. Based on the original data, the root-mean-square difference is calculated as:

$$w_0 = \sqrt{\frac{\sum_{i=1}^{9} w_i^2}{9}} = 0.828. \tag{6.43}$$

In the analysis of the initial differences, the absolute difference $w_6 = 1.98$ draws attention, as it significantly deviates from all other results.

To test the hypothesis that this value represents a statistical outlier in the laboratory with $w_6 = 1.98$, we apply Cochran's test:

$$G = \frac{\max\left(S_i^2\right)}{\sum\limits_{i=1}^{n} S_i^2} = \frac{1.98^2}{6.1663} = 0.636. \tag{6.44}$$

We refer to the G-distribution table, from which we find that the critical values are as follows: $G_{crit}(5\%) = 0.638$ and $G_{crit}(10\%) = 0.754$.

Thus, the difference $w_6 = 1.98$ can be considered a quasi-outlier. If a traditional approach is applied for the statistical processing of the data, the value $w_6 = 1.98$ should be excluded from the observations. In this case, the resulting "refined" ($n = 8$) root-mean-square deviation $w_0 = 0.53$ differs by approximately 1.5 times from the value calculated over the full set of initial data ($n = 9$). As can be seen, excluding values from the original data that lie along the line separating quasi-outliers significantly affects the outcome of the analysis—the estimate of the repeatability standard deviation of the testing method.

TABLE 6.4
Application of the "S-Algorithm" for Determining Differences between Laboratories

Iteration j	0	1	2	3	4	5
ψ_j	-	0.66	0.86	1.00	1.09	1.12
$w_1{}^*$	0.00	0.00	0.00	0.00	0.00	0.00
$w_2{}^*$	0.28	0.28	0.28	0.28	0.28	0.28
$w_3{}^*$	0.32	0.32	0.32	0.32	0.32	0.32
$w_4{}^*$	0.35	0.35	0.35	0.35	0.35	0.35
$w_5{}^*$	0.40	0.40	0.40	0.40	0.40	0.40
$w_6{}^*$	0.49	0.49	0.49	0.49	0.49	0.49
$w_7{}^*$	0.80	0.66	0.80	0.80	0.80	0.80
$w_8{}^*$	0.95	0.66	0.86	0.95	0.95	0.95
$w_9{}^*$	1.98	0.66	0.86	1.00	1.09	1.12
Root-mean-square deviation $w_j{}^*$	0.83	0.47	0.56	0.60	0.62	0.62
Updated value w^*	0.40	0.52	0.61	0.66	0.68	0.68

As previously mentioned, the use of robust methods (in this case, the "S-algorithm") allows the estimation of the repeatability standard deviation based on the results obtained from all $n = 9$ laboratories. First, we order the laboratories according to the increase in absolute differences. In Table 6.4, the indices of the differences in ascending order are marked with an asterisk.

Since the number of degrees of freedom is $v = 1$, we select the values of the consistency factor $\xi = 1.097$ and the limiting factor $\eta = 1.645$ from Table 6.3.

We then proceed to the iterative procedure.

In the first step, we determine: $\psi_1 = w_5{}^*\eta = 0.4 \cdot 1.645 = 0.66$.

Thus, the limiting value at the first step of the robust procedure is $\psi_1 = 0.66$.

We perform a modification of the absolute differences using the following correspondence:

$$w_{ij}^* = \begin{cases} \psi_j \; if \; w_i > \psi_j, \\ w_i \text{ in other cases,} \end{cases} \quad j = 0, 1, \ldots \tag{6.45}$$

The values w_i from the original data must be modified, as they exceed ψ_1. As a result, we obtain a column of modified data at the first step of the iteration.

Using the modified differences obtained in the first iteration step w_{i1}^*, we calculate the unbiased (robust) estimate of the root-mean-square deviation:

$$w_1^* = \xi \sqrt{\frac{\sum_{i=1}^{n} \left(w_{i1}^*\right)^2}{n}} = 1.097 \cdot 0.47 = 0.52. \tag{6.46}$$

Thus, the "Updated value" limiting value, which the random variable does not exceed with probability $\alpha = 0.1$, is given by: $\psi_2 = 1.645 \cdot 0.52 = 0.86$.

Thus, as a result of the first iteration, the value of $w_7{}^*$ has been "brought within" the prescribed limiting value.

For the eighth and ninth results, the modified values will be: $w_{82}{}^* = 0.86$ and $w_{92}{}^* = 0.86$.

The modified values for the second iteration step are presented in accordance with Table 6.4.

We calculate the value $w_2 = 0.56$ and then determine the refined root-mean-square deviation: $w_3{}^* = \xi w_3 = 1.097 \cdot 0.56 = 0.61$. And the limited (threshold) value: $\psi_3 = 1.645 \cdot w_3{}^* = 1.0$.

We proceed to the third iteration. Since the value of $w_{80}{}^* < \psi 3$ is below the threshold, it does not need to be modified. According to expression (6.45), only $w_{90}{}^* = 1$ is modified.

In the fourth iteration step, we obtain a robust value $w_4{}^* = 0.68$, which differs from the previous value $w_3{}^* = 0.66$ by $(0.02/0.66)100\% = 3\%$.

As a consequence, from Table 6.4, row 4, only $w_9{}^* = 1.98$ exceeds ψ_4. Continuing the iterative procedure (fifth step), aimed at increasing the estimate, shows that the root-mean-square deviation is refined only in the third decimal place.

Thus, the robust root-mean-square difference is accepted as $w^* = 0.68$. It lies between the two values: $w = 0.83$, calculated using all laboratory differences including the quasi-outlier, and $w' = 0.53$, calculated using eight values excluding the quasi-outlier. Therefore, the statistically reliable estimate of the standard deviation of repeatability of the tested method is:

$$\sigma_r = \frac{w^*}{\sqrt{2}} = \frac{0.68}{\sqrt{2}} = 0.48. \tag{6.47}$$

Thus, robust methods, in particular the "S-algorithm," allow the determination of the standard deviation for a normally distributed variable that best corresponds to the obtained experimental data. Since all experimental data were used, this estimate possesses the maximum possible statistical reliability.

6.6 METHODOLOGY FOR USING THE ABSOLUTE MEDIAN DEVIATION IN THE ASSESSMENT OF TEST RESULT ACCURACY

As previously mentioned, when assessing the accuracy characteristics of a testing method, a collaborative experiment is conducted, involving laboratories with appropriate professional training. In such interlaboratory experiments, it is assumed that all participating laboratory personnel have the same level of expertise. In practice, however, it is often found that some laboratories have poorer organization of testing procedures, due to objective reasons. According to expression (6.26), when determining reproducibility, the Grubbs' test is applied to the laboratory mean results and their associated scatter, in order to exclude so-called outliers. Excluding results from these laboratories, on the one hand, slightly idealizes the actual testing conditions, and on the other hand, increases the uncertainty of the resulting estimate.

All statistical processing and decisions based on it rely on the assumption of normality of the distribution, which is generally justified because a well-developed theory of statistical inference exists. However, in many practical cases, there is insufficient data to construct parametric models that adequately represent the experimental results. As early as the 1960s, prominent statisticians, based on detailed experiments, concluded that data processed under the assumption of normality typically contain, on average, 10% gross errors (ranging from 1% to 20%), either overtly or covertly.

All theoretical assumptions are based on the possibility of conducting experiments (observations) under identical, unchanging conditions. In practice, there is spatiotemporal variability in the observation conditions, including variability of the test object itself. Consequently, one can agree with Tukey's statement that "normality is a myth, and normal distributions have never existed and never will." Causes of outliers in experimental results can include operator errors, failures of measurement instruments used in testing, or deviations from experimental procedures.

The contamination model is characterized by extended "tails" of the probability density. In the contamination scheme, there exists a central robust part of the distribution, determined by ordinary small error components, and extended tails, representing relatively rare outliers. Using this scheme allows one, on the one hand, to maintain a convenient probabilistic assumption of homogeneity of the hypothetical population, on which all probabilistic estimates are based, and on the other hand, to introduce the possibility of high-level errors.

It is known that, due to the influence of random variables, one can determine an estimate of the true value of a parameter with an error. Typically, errors relative to the true value are asymmetrically distributed, and smaller sample sizes exhibit greater asymmetry. The sample mean minimizes the sum of squares of deviations from the true value, but this is the source of its sensitivity to outliers, which can significantly increase the sum.

There also exists a central part of the experimental data distribution, corresponding to the assumed population distribution. For data within this central region, it is appropriate to perform least squares method (LSM). The Laplace modular criterion is more robust to outliers than LSM, providing the best results under highly unfavorable distributions. Therefore, based on the above reasoning, in constructing robust methods a "hybrid approach" is formed: for a central group, the LSM is applied, while beyond a certain threshold, to reduce the influence of outliers while preserving data, the modular criterion is used.

To reduce sensitivity to outliers, it is proposed to minimize the functional:

$$\sum_{i=1}^{n} \rho(x_i - \mu), \tag{6.48}$$

where $\rho(\varepsilon)$ is the loss function, which allows for a less "strict" approach to rejecting outliers that deviate from the center of the distribution by value $\varepsilon = x_i - \mu$ greater than $c\sigma$.

The constant c controls the degree of robustness, and its value depends on the level of contamination. For instance, for 1% contamination $c = 2$, while for 5% contamination $c = 1.4$. Typically, $c = 1.5$ is chosen.

According to the criteria selected, the existing data should be modified as follows:

$$x_i^* = \begin{cases} x_i \text{ at } |x_i - \widehat{\mu}| \le c\sigma, \\ \widehat{\mu} - c\sigma \text{ at } x_i < \widehat{\mu} - c\sigma, \\ \widehat{\mu} + c\sigma \text{ at } x_i > \widehat{\mu} + c\sigma, \end{cases} \quad (6.49)$$

where $x_i^* = \text{med}\{x_i\}$ (x_i preliminarily ranked in ascending order).

Experiments have shown that the midpoint of the interval between the sample quartiles possesses the best properties in terms of robustness against outliers. Therefore, as the initial estimate of the distribution center $\widehat{\mu}$, which is resistant to outliers, the sample median $\text{med}\{x_i\}$ is used. As a robust measure of spread, the interquartile range (IQR) is taken—that is, the difference between the values of the third quartile ($f(x) = 3/4$) and the first quartile ($f(x) = 1/4$).

The IQR is a measure of distribution spread. The area under the probability density curve within this interval (see Figure 6.3) accounts for 50% of the total probability. Assuming a particular probability distribution law, the length of this interval uniquely corresponds to the variance of that distribution.

To relate the IQR to the variance, the absolute median deviation is introduced, which serves as an estimate of the scale—effectively transforming the "full" distribution into the "adjusted" one:

$$MAD_n = \text{med}\{|x_i - M_n|\}, \quad (6.50)$$

where $M_n = \text{med}\{x_i\}$, x_i is a sample element, and the index n corresponds to the number of elements in the sample.

To establish the relationship between the parameters of the "truncated" distribution and the assumed population distribution, that is, to perform the scaling condition, it is necessary to recalculate the standard deviation σ using a correction factor determined from the original probability density function:

$$f(x) = \frac{d}{\sigma\sqrt{2\pi}} e^{-\frac{(x-\mu)^2}{2\sigma^2}}. \quad (6.51)$$

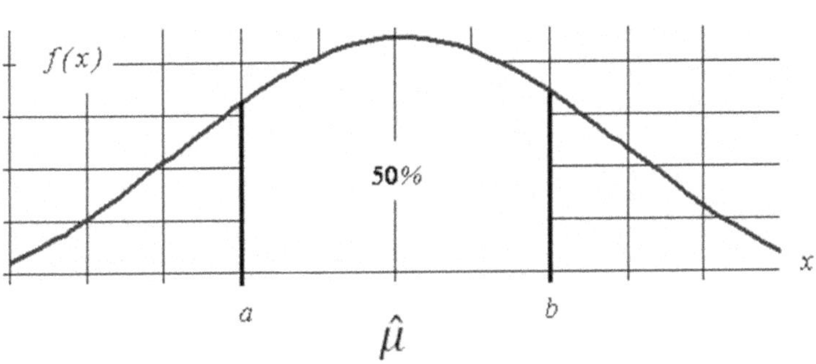

FIGURE 6.3 Determination of the interquartile range: a and b correspond to the ordinates of the first and third quartiles.

For the "non-truncated" distribution $d = 1$, for the interquartile "truncated" distribution:

$$d = \frac{0.5}{F\left(\frac{b-\mu}{\sigma}\right) - F\left(\frac{a-\mu}{\sigma}\right)}, \tag{6.52}$$

where $F(z)$ is the cumulative distribution function of the standard normal distribution.

Using the standard normal distribution tables, the value of $d = \frac{1}{0.6745} = 1.483$.

Based on this, the scale adjustment factor when transitioning from the "non-truncated" distribution to the "truncated" one is 1.483.

Thus, the estimate of the standard deviation S^* for a sample of size n is determined on the basis of the normalized IQR and is given by:

$$S^* = 1.483MAD_n. \tag{6.53}$$

In comparison with the estimate of the standard deviation determined based on the well-known expression:

$$S = \sqrt{\frac{\sum_{i=1}^{n}(x_i - \overline{x})^2}{n-1}}, \tag{6.54}$$

it will remain robust even if the sample contains up to 50% of anomalous results.

Accordingly, to implement the outlier-robustness criterion based on the selected loss function, it is necessary to determine the median of the initial data set. Then, using expression (6.53), calculate the value S^*, which, together with the chosen value $c = 1.5$, will make it possible to establish the initial threshold for the transition from the LSM to the least absolute deviations method, namely:

$$\varphi = cS^*. \tag{6.55}$$

As the distribution estimate, it is necessary to use the median of the ordered data set, which is more robust to outliers than the mean value.

By applying condition (6.49), we obtain the modified data series x_i^*, for which the mean value is then calculated:

$$\overline{x}^* = \sum_{i=1}^{n} \frac{x_i^*}{n}. \tag{6.56}$$

Additionally, according to expression (6.54), after substituting x_i^* and $\overline{x}^*$, the standard deviation $S(x^*)$ is determined.

The calculated value $S(x^*)$ is then used to compute the "updated" robust standard deviation:

$$S^* = 1.134S(x^*), \tag{6.57}$$

where the coefficient 1.134 corresponds to $c = 1.5$ for a normal distribution.

Coefficient 1.134 accounts for the fact that the data are analyzed for a truncated distribution with $c = 1.5$, which corresponds to 86.6% of the assumed normal distribution, to which the ascending data—including anomalous results—are fitted. This fitting is performed sequentially in several stages, at each of which the input values are the modified data according to (6.49).

The obtained value S^* is then used to calculate the updated value for the next (or previous) iteration stage.

The convergence of the algorithm is determined by the difference between the values of $\bar{x}^*$ calculated at the current and next (or previous) iteration steps. The procedure is repeated until $\bar{x}^*$ and S^* values from one calculation to the next become minimal.

6.7 EXAMPLE OF USING THE ABSOLUTE MEDIAN DEVIATION IN THE ASSESSMENT OF REPRODUCIBILITY OF A LABORATORY TESTING METHOD

Collaborative laboratory studies were conducted with the aim of determining the accuracy characteristics of the testing method. For this purpose, $n = 9$ laboratories were involved. Based on the results of the studies, the following mean values were obtained in the laboratories:

24.140; 20.155; 19.500; 20.300; 20.705; 17.570; 20.100; 20.940; 21.185.

In Table 6.5, the column labeled "0" presents the ordered ascending data ($j = 0$, initial iteration) $x_{i(0)}$, $i = 1, \ldots, 9$. Based on these data, the mean value $x_0^* = 20.511$ is calculated, and for $n = 9$ using expression (6.54), the standard deviation estimate is $S_0^* = 1.727$.

As the initial estimate of the distribution center for the population, the median $x_5^* = 20.3$ is taken and using expression (6.53), the standard deviation estimate based on the IQR is determined and used as the updated initial value of the standard deviation $S_1^* = 0.949$.

Next, for the first iteration step ($j = 1$), the value $\varphi_1 = 1.5S_1^* = 1.434$ is calculated, which is then used to compute the threshold limits:

$$x_{j\ \min}^* = x_{j-1}^* - \varphi_1 = 18.876 \text{ and } x_{j\ \max}^* = x_{j-1}^* + \varphi_1 = 21.724. \quad (6.58)$$

Based on the previously obtained limits, calculations are performed for the jth column using expression (6.49). These data are then used to compute the value of S_j^* according to (6.57).

From the obtained value S_j^*, the next iteration value φ_{j+1} is calculated, and so on.

The results of the calculations are entered into the corresponding column of Table 6.5. The algorithm is then applied iteratively, as previously described. As can be seen from the results presented in Table 6.5, the mean values calculated at the fourth $x_4^* = 20.411$ and fifth $x_5^* = 20.412$ iteration steps differ negligibly.

TABLE 6.5
Application of the Method to Mean Values in the Elements (% of Crosstalk)

Iteration	0	1	2	3	4
Φ		1.424	1.478	1.514	1.539
$x^* - \varphi$		18.876	18.909	18.893	18.872
$x^* + \varphi$		21.724	21.865	21.921	21.95
x_1^*	17.57	18.876	18.909	18.893	18.872
x_2^*	19.5	19.5	19.5	19.5	19.5
x_3^*	20.1	20.1	20.1	20.1	20.1
x_4^*	20.155	20.155	20.155	20.155	20.155
x_5^*	20.3	20.3	20.3	20.3	20.3
x_6^*	20.705	20.705	20.705	20.705	20.705
x_7^*	20.94	20.94	20.94	20.94	20.94
x_8^*	21.185	21.185	21.185	21.185	21.185
x_9^*	24.14	21.724	21.865	21.921	21.95
Mean values	20.511	20.387	20.407	20.411	20.412
Standard deviations	1.727	0.869	0.89	0.905	0.916
Updated estimates x^*	20.3	20.387	20.407	20.411	20.412
Updated estimates S^*	0.949	0.985	1.009	1.026	1.039

Meanwhile, the robust estimate of the standard deviation of the mean values obtained across the laboratories corresponds to $S_4^* = 1.026$ and $S_5^* = 1.039$.

If the traditional approach is used, that is, detecting outliers with Grubbs' test, and the outliers from laboratories $x_{1(0)}^* = 17.57$ and $x_{9(0)}^* = 24.14$ are excluded, then for $n = 7$ laboratories the mean value is $\mu = 20.412$. This indicates a convergence of the distribution centers of the sample values, that is, the same estimate of the accuracy characteristic of the tested method is obtained.

However, applying the robust method yields a value $S^* = 1.039$ that reflects the true dispersion of the original measurements. Excluding the two outliers (the mean values from the "first" and "ninth" laboratories) from the results of the collaborative experimental tests gives $S = 0.501$, which deviates from the actual experimental situation, effectively idealizing it and assuming that the experimental data must belong to a population with a normal distribution. Since only $n = 7$ laboratories were considered in this case, the obtained standard deviation estimate has lower statistical reliability. The discrepancy between the obtained standard deviation estimates can also be used to analyze the organization of the testing process in the two laboratories.

Notes

1 ISO 5725-1:2023, *Accuracy (trueness and precision) of measurement methods and results — Part 1: General principles and definitions* (International Organization for Standardization, 2023).

2 Semyon G. Rabinovich, *Evaluating measurement accuracy: a practical approach*: [book], Springer International Publishing, 2017, 323 p.

3 Dennis Wackerly, William Mendenhall, Richard Scheaffer, *Mathematical statistics with applications*: [book], Thomson Brooks/Cole, 2008, 944 p.
4 B. N. Taylor and C. E. Kuyatt, *Guidelines for evaluating and expressing the uncertainty of NIST measurement results*, NIST Technical Note 1297, Rev. ed., National Institute of Standards and Technology, Gaithersburg, MD, 2008.
5 ISO/IEC Guide 98-1:2024, *Guide to the expression of uncertainty in measurement* (Int. Organ. Stand., Geneva, 2024: 13).
6 ISO/TS 21748:2004, *Guidance for the use of repeatability, reproducibility and trueness estimates in measurement uncertainty estimation* (International Organization for Standardization, Geneva, 2004).
7 ISO 5725-5:1998/Cor 1:2005, *Accuracy (trueness and precision) of measurement methods and results — Part 5: Alternative methods for the determination of the precision of a standard measurement method* (International Organization for Standardization, 2005).

BIBLIOGRAPHY

ISO 5725-1:2023, *Accuracy (trueness and precision) of measurement methods and results — Part 1: General principles and definitions*, International Organization for Standardization (2023).

ISO 5725-5:1998/Cor 1:2005, *Accuracy (trueness and precision) of measurement methods and results — Part 5: Alternative methods for the determination of the precision of a standard measurement method*, International Organization for Standardization (2005).

ISO/IEC Guide 98-1:2024, *Guide to the expression of uncertainty in measurement*, Int. Organ. Stand., Geneva (2024): 13.

ISO/TS 21748:2004, *Guidance for the use of repeatability, reproducibility and trueness estimates in measurement uncertainty estimation*, International Organization for Standardization, Geneva (2004).

Rabinovich, Semyon G., *Evaluating measurement accuracy: A practical approach*: [book], Springer International Publishing (2017), 323 p., ISBN 9783319601243.

Taylor, B. N. and C. E. Kuyatt, *Guidelines for evaluating and expressing the uncertainty of NIST measurement results*, NIST Technical Note 1297, Rev. ed., National Institute of Standards and Technology, Gaithersburg, MD (2008). https://doi.org/10.6028/NIST.TN.1297

Wackerly, Dennis, William Mendenhall, and Richard Scheaffer, *Mathematical statistics with applications*: [book], Thomson Brooks/Cole (2008), 944 p., ISBN 978-0495110811.

Index

Pages in *italics* refer to figures and pages in **bold** refer to tables

M

N

O

P

R

S

T

U

V

For Product Safety Concerns and Information please contact our EU representative GPSR@taylorandfrancis.com
Taylor & Francis Verlag GmbH, Kaufingerstraße 24, 80331 München, Germany

www.ingramcontent.com/pod-product-compliance
Ingram Content Group UK Ltd.
Pitfield, Milton Keynes, MK11 3LW, UK
UKHW020201200926
475702UK00004B/566

* 9 7 8 1 0 4 1 2 4 5 9 9 5 *